GEOTECHNICAL SPECIAL

MW00717352

GROUTS AND GROUTING

A Potpourri of Projects

PROCEEDINGS OF SESSIONS OF GEO-CONGRESS 98

SPONSORED BY
The Grouting Committee of
The Geo-Institute
of the American Society of Civil Engineers

October 18–21, 1998
Boston, Massachusetts

EDITED BY
Larry Johnsen
Dick Berry

1801 ALEXANDER BELL DRIVE
RESTON, VIRGINIA 20191–4400

Abstract: This proceedings, *Grouts and Grouting: A Potpourri of Projects*, consists of papers presented at sessions sponsored by the Geo-Institute in conjunction with the ASCE Annual Convention held in Boston, Massachusetts, October 18-21, 1998. These papers discuss unusual case studies where grouting was used to address widely different situations. Projects involve mines, tunnels, dams, walls, sinkholes, historic buildings, river banks, and earthquakes. The types of grouting involved are jet, displacement, and permeation, while the types of grout are cement, chemical, and microfine cement. Unusual monitoring and control methods are involved in some of the case studies.

Library of Congress Cataloging-in-Publication Data

Grouts and grouting: a potpourri of projects: proceedings of sessions sponsored by the Grouting Committee of the Geo-Institute of the American Society of Civil Engineers, Boston, Massachusetts, October 18-21, 1998 / edited by Larry Johnsen and Dick Berry.
    p.   cm.   –(Geotechnical special publication; no. 80)
    Includes bibliographical references and index.
    ISBN 0-7844-0386-4
1. Grouting–Congresses. 2. Grout (Mortar)–Congresses. I. Johnsen, Larry. II. Berry, Richard M. III. American Society of Civil Engineers. Geo-Institute. Grouting Committee. IV. Series.
    TA755.G77   1998   98-39146
    693'.1–dc21   CIP

Geotechnical Special Publications

PREFACE

These grouting sessions were envisioned as rather a potpourri of projects offering unusual perspectives and project approaches to the use of grouting for the greater variety of applications which are emerging in the United States. As the country builds more and more on marginal land and constricted sites, the use of grouting to improve those sites and build over older construction is causing a revival of sorts in grouting over our long history when most new construction went to new and better land in the suburbs.

Papers run from one about sand remediation under a settling wall which should have been improved before construction in Florida to scour remediation in Illinois - from abutment stabilization in Pennsylvania to a salt mine closure in Detroit - from jet grouting in a New York tunnel to grouting during rock tunneling in California.

For those who wish to add and comment, all the papers are open for discussion in the ASCE Journal of Geotechnical and Geoenvironmental Engineering. Let us hear from you.

The editors want to thank the following reviewers and referees of papers for their assistance. Without the capable assistance of these gentlemen, most of these papers would not have made it to their timely publication.

Peter P. Aberle, Consultant, Lakewood, CO
Robert Alperstein, Principal, R A Consultants, Wayne, NJ
Dr. Stanley M. Bemben, Consultant, New Britain, CT
Dr. Roy H. Borden, Prof., North Carolina State, Raleigh, NC
Dr. Donald A. Bruce, Principal, ECO Geosystems, Venetia, PA
Michael J. Byle, Manager, Gannett Fleming, King of Prussia, PA
Eric R. Drooff, Manager, Hayward Baker, Odenton, MD
Dr. James P. Gould, Principal, Mueser Rutledge, New York, NY
Steven D. Scherer, VP, TCDI, Lincolnshire, IL
James Warner, Consultant, Mariposa, CA
Kenneth D. Weaver, Consultant, Fremont, CA
Joseph P. Welsh, VP, Hayward Baker, Odenton, MD
Dr. Peter T. Yen, Manager, Bechtel Corp., San Francisco, CA

Editors and Reviewers

Richard M. Berry, Chairman, Rembco Engineering, Knoxville, TN
Lawrence F. Johnsen, Principal, Heller and Johnsen, Stratford, CT

Contents

A Jet Grout Stabilized Excavation Beneath An Existing Building

Chu Eu Ho[1], MASCE

Abstract

A new single basement was constructed as part of a building refurbishment project in Singapore. The excavation was carried out within the tight confines of the existing building. The sides and bottom of the excavation were stabilized by jet grout columns installed using the triple-tube technique. The excavation was carried out without shoring. The installation of the jet grouting works and subsequent excavation was monitored by geotechnical instruments. This paper presents the monitoring data obtained and discusses the influence of the jet grouting and construction activities on the surrounding ground and structures.

Introduction

The refurbishment of the Commercial Union House at Robinson Road involved the construction of a new single level basement to create additional space for mechanical and electrical plant room facilities. The project was located in the busy Central Business District in downtown Singapore. The site was surrounded by major roads and a backlane. The eastbound and westbound tunnels of the Mass Rapid Transit (MRT) system ran within 10m of the site boundary beneath Robinson Road, Fig.1. The development of the new basement fell within the Second Reserve of the MRT Protection Zone and had to comply with strict constraints of MRTC Code of Practice. Fig.2 shows the proximity of the tunnels in relation to the excavation. The tunnels were 6m in diameter and located one above the other. The crown of the upper and lower tunnels were approximately 9m and 18m below the street level respectively.

---

[1] General Manager, Ground Engineering Division, Presscrete Engineering Pte Ltd, 31 Changi South Avenue 2, Singapore 486478, Singapore.

1

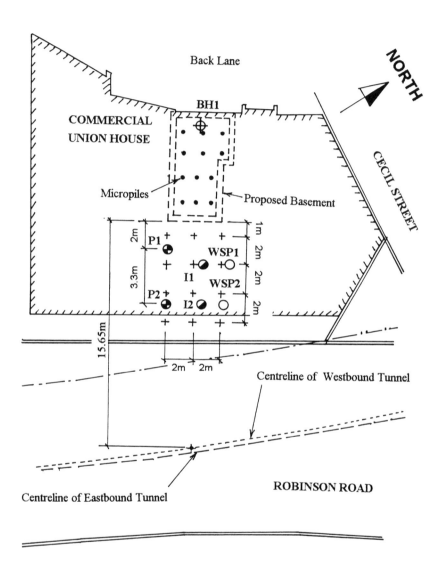

Fig. 1 Site Plan

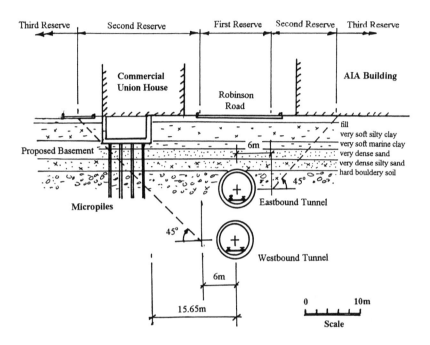

Fig. 2 Proximity of Mass Rapit Transit Tunnels

The size of the basement to be constructed was about 7.8m by 4.5m in plan and 3.5m deep. Micropiles were required to resist the uplift water pressure at minimal dead load condition. The construction had to be carried out within the tight confines of the existing first storey area.

Table 1 Subsoil profile and geotechnical properties

Depth 0 to 0.3m
Concrete slab

Depth 0.3m to 1.2m Backfilled soil, mixture of sand, silty clay and concrete fragments	N = 2 @ 1.8m	

Depth 1.2m to 4.2m
Very soft yellow, brown, reddish brown
silty clay with fine to coarse grained sand   N = 4 @ 3.9m

Sample from 3.0m to 3.6m
$Cu = 23$ kPa $\phi_u = 1°$ $\gamma_b = 19.5$ kN/m3
MC = 27%   LL = 56%   PL = 23%
$C' = 10$ kPa $\phi' = 28°$ $\gamma_b = 19.9$ kN/m3

Depth 4.2m to 5.6m
Mixture of gravel size corals and   N = 1 @ 4.8m
very soft bluish grey marine clay

Depth 5.6m to 7.2m
Very dense grey clayey sand with shell   N = 100/0.2m @ 6.25m
particles and gravel size coral fragments

Depth 7.2m to 9.6m
Medium dense to very dense reddish   N = 24 @ 7.8m
brown yellowish brown fine to medium
grained silty sand with shell fragments   N = 79 @ 9.3m
and some gravels with rock fragments

Depth 9.6m to 19.6m
Bouldery soil; weak to moderately strong   CR = 57% RQD = 0   from 9.6m to 10.6m
moderately to highly weathered very poor   CR = 90% RQD = 0   from 10.6m to 11.6m
fractured light greyish white siltstone   CR = 85% RQD = 0   from 11.6m to 12.6m
boulders and silty clay mixture   CR = 58% RQD = 0   from 12.6m to 13.6m
                                            CR = 52% RQD = 0   from 13.6m to 14.6m
                                            CR = 62% RQD = 0   from 14.6m to 15.6m
                                            CR = 76% RQD = 0   from 15.6m to 16.6m
                                            CR = 54% RQD = 0   from 16.6m to 17.6m
                                            CR = 68% RQD = 11% from 17.6m to 18.6m
                                            CR = 55% RQD = 0   from 18.6m to 19.6m

---

Site Geology

A single borehole (BH1) was sunk at the location of the proposed basement to a depth of 19.6m to establish the subsoil conditions (Fig1). The soil investigation revealed that the site was underlain by old reclamation fill followed by marine clay and sand layers. Underlying these was a conglomerate comprising stiff to hard clayey silt or silty clay matrix with random inclusions of siltstone boulders of differing weathering grades. The existing ground level was about +102.3mOD and the water table was located between 0.9m and 1.02m below ground. Standard Penetration Tests were carried out in the soils above the bouldery soil. Within the bouldery layer, continuous cores were taken for determination of Core Recovery (CR) and Rock Quality Designation (RQD) indices. An undisturbed sample was taken in the marine clay for measurements of undrained (C_u, ϕ_u) and drained (C', ϕ') shear strengths, as well as for bulk density (γ_b), moisture content (MC), liquid limit (LL) and plastic limit (PL). The subsoil profile and geotechnical properties obtained from borehole BH1 are summarised in Table 1.

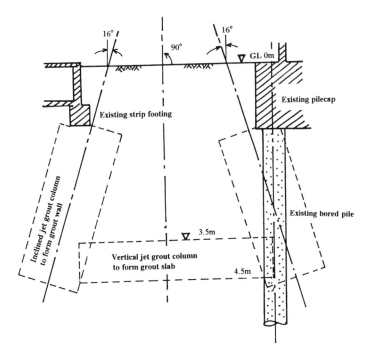

Fig. 3 Jet grout columns for side wall and bottom slab

Jet Grout Scheme

Due to the constraints of the existing building structure and low headroom between floors, a normal retaining wall and bracing support system was considered to be impractical. A jet grout scheme was therefore adopted to enable the excavation for the basement to be carried out without shoring. Fig.3 shows a typical cross-section of the excavation pit. The jet grout scheme involved the installation of 1.6m diameter overlapping jet grout columns to form a grout wall for retention of the excavation sides. A 1m thick continuous grout slab was also formed from jet grout columns to provide a bottom seal and also to act as a buried strut. Spacing for the vertical columns were kept at 1.4m x 1.4m grids or closer. Wherever possible the jet grout columns were installed in a vertical direction. However, there were cases where obstruction was posed by existing structural walls which could not be removed. At these areas, the jet grout columns were installed at an inclined direction of 16° to the vertical beneath the existing structures. For most of the sides of the pit the jet grout columns were formed directly beneath existing footings and ground slabs. Along one particular stretch, the jet grout columns had to be installed immediately next to existing bored piles. Fig.4 shows the layout plan of the jet grout columns The grout wall was formed from 8 nos. vertical columns 5.25m deep (W1 to W8) and 17 nos. inclined columns 4.5m deep (W9 to W25). The bottom grout slab was formed from 23 nos. vertical columns 1m deep (B1 to B23).

Jet Grout Trial

Due to the sensitivtiy of the existing building and the proximity of the MRT tunnels, it was decided that the triple-tube jet grouting system be adopted for installing the jet grout columns. A preliminary trial was conducted at the location of one jet grout column (W5) closest to the MRT tunnels on 28 July 1994 to study the effects of the grouting activity (Fig.4). Prior to the trial, inclinometers, water standpipes and pneumatic piezometers were installed in the ground to monitor the displacements and pore pressure build-up in the soils. The installation depths of inclinometers I1 and I2 were 14.5m and 15m respectively. Water standpipes WSP1 and WSP2 were both installed to 6m deep. The tips of the piezometers P1 and P2 were located at 6m depth within the clayey sand layer immediately beneath the marine clay. Initial readings for all instruments were taken on 29 June 1994. Survey markers were also implanted in the ground and existing structural columns of the building for measurement of settlement or heave. Datum levels were established on 7 June 1994. Fig.1 shows the layout of the instrumentation.

For the trial, the grout column was installed up to within 0.5m of the ground surface. The drilling was carried out using the wet drilling technique with a drag

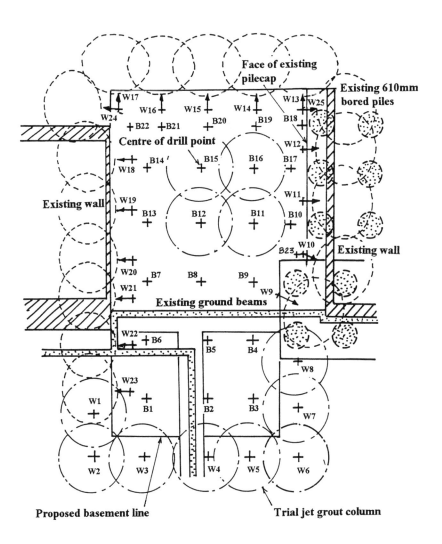

Fig. 4 Layout of jet grout columns

bit to form a 120mm diameter bore. No casing was used to support the drillhole. The parameters adopted for the jet grouting operation are given in Table2.

Table 2 Summary of Trial Parameters

Drilling time	1500 to 1545 hrs
Grouting time	1630 to 1800 hrs
Withdrawal speed	12 min/m
Rotational speed	6 rpm
Water jet pressure	400 bars
Water discharge rate	100 l/min
Grout pressure	30 to 40 bars
Grout discharge rate	70 l/min
Air pressure	7 bars
Water-cement ratio	70 kg : 50 kg

There were no significant heave or settlement caused by the jet grouting trial. Level surveys indicated that ground surface changes were of the order of +1mm or -1mm. No obvious displacement of the structural columns were noticed. The maximum resultant lateral displacement of inclinometers I1 and I2 located at 3m and 5.3m from the trial jetting position was 6.02mm and 3.18mm respectively, recorded 17 hours after the trial was completed (Fig.5). These displacements took place within the soft silty clay and marine clay layers and were in the direction away from the jet grouting activity (Fig.6 and 7). Readings taken on the 39th day after the trial (5 Sep 94) indicated a further increase to 6.97mm for I1 and a decrease to 2.77mm for I2. The following observations were also made from the monitoring data after one day. Changes in water levels were significant, with rises of 0.35m and 0.21m being recorded for WSP1 and WSP2 respectively (Fig.8). Some air bubbles were observed to have escaped from crevices in the ground slabs. The rise in water levels appeared to have been caused by the generation of pressurised flow through the relatively porous backfill soil immediately beneath the existing structural slabs. There was however a small reduction of 2.4% and 6.3% in piezometric pressures from the pre-trial values of 4.1m (P1) and 4.8m (P2) respectively (Fig.9).

Cores were taken from the trial jet grout column on 22 August 1994. The stress-strain curve obtained from an unconfined compression test on a typical 99.5mm diameter x 199.5mm height grout core after 28 days was approximately linear. The maximum deviator stress was 1576 kPa at the maximum strain of 0.12%. The computed grout strength and elastic modulus were therefore 6.03MPa and 1.313GPa respectively. Density of the grout sample was 15.6 kN/m3. The results were therefore satisfactory.

Installation of Jet Grout Columns

The jet grout columns subsequent to the trial were installed between 20 September 1994 and 23 October 1994 over a period of approximately one month with a maximum production rate of 2 columns a day. The jet grout columns nearest to the tunnels were executed first to form an initial cut-off wall against potential influence of the installation activities for subsequent jet grout columns further behind. W5 the trial column had already been formed. The installation sequence for the cut-off wall was W1, W3, W6, W8, W23, W2, W4, W7 and W22. Next the columns forming the grout slab abutting the cut-off wall were

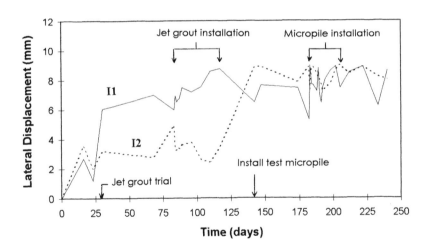

Fig. 5 Maximum resultant lateral ground displacements

executed in the following sequence B1, B3, B5, B6, B2 and B4. This sequence ensures that the stiffness of the ground facing the tunnels was always increased with each additional installation of a new jet grout column further behind. Any potential displacement of the soft clays due to blockage of the return sludge flowing up the drilled hole would result in a deflection of the soft clay in the direction away from the tunnels. The next step involved the installation of inclined jet grout columns underneath and immediately adjacent to existing footings, pilecaps and bored piles to form the remaining portions of the grout wall. The sequence adopted was W9, W11, W13, W15, W17, W10, W12, W14, W16, W18, W20, W19, W21, W24 and W25. Again it was necessary to install the jet grout columns nearest to the foundations first prior to those for the internal grout slab to ensure minimum disturbance to existing structures. The final sequence for jet grout installation was B9, B12, B10, B8, B11, B13, B15, B18, B7, B14, B16, B20, B22, B17, B19, B21 and B23 to complete the grout slab.

The maximum observed resultant lateral displacements induced by jet grouting activity were 8.77mm for I1 (24 Oct 94) and 3.79mm for I2 (3 Oct 94). Fig.5 shows that the lateral displacements at I1, which was nearest to the jet grouting activity, was generally accumulative for each successive execution of the jet grout columns. Since the recommencement of jet grout installation on 20 Sep 94 after the trial, I1 registered a further displacement of 2.8mm upon completion of grouting works. The displacement trend was not so obvious for I2. In both cases, the maximum displacement profiles were located at 3m to 4m depth within the soft clays. After all grouting works were completed, it was noted that there was no significant rebound of the ground. This implies that the induced ground displacements were plastic in nature and hence irreversible. These observations were also reported by Ho (1995) for jet grouting works in soft marine clay at the Singapore Post Centre site.

Water standpipe readings (Fig.8) indicated that there was significant change to the ground water regime due to jet grouting activity, with increase in levels of 0.59m to 0.62m between 20 Sep 94 to 3 Oct 94. After this time, it was clearly demonstrated that once the initial cut-off grout wall was formed, the water levels started to drop to a more intermediate value, implying that the effects of the jet grouting activity had been contained. Fig.9 shows that there was a small temporary build up in pore pressures of 0.5m to 0.6m head of water in the clayey sand layer within the same period. The rise and fall of the piezometric readings follow closely the trend established by the water level measurements discussed above. The screening effect of the cut-off grout wall was therefore quite evident.

Over the period of jet grouting works, survey monitoring of immediate structural columns of the building indicated that settlements of the order of 3mm had been induced. These movements were attributable to subsidence occurring in the sandy backfill just beneath the ground. The high pressure water jet used for

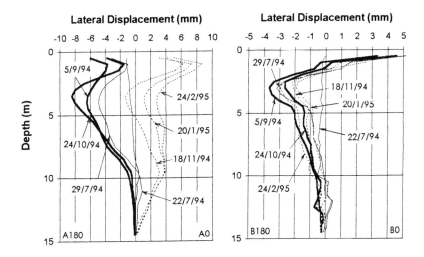

Fig. 6 Observed displacements in inclinometer I1

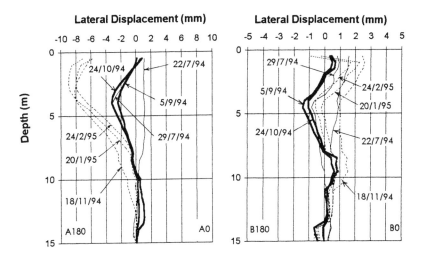

Fig. 7 Observed displacements in inclinometer I2

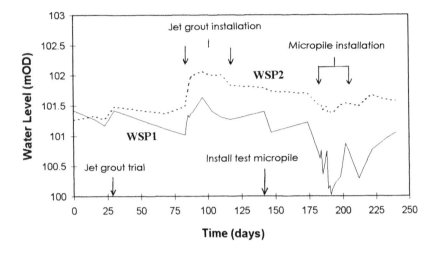

Fig. 8 Observed water table levels

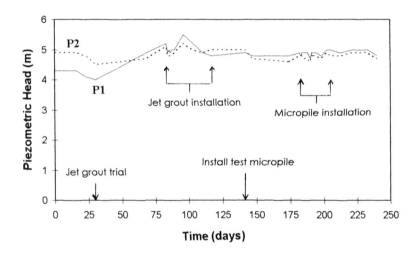

Fig. 9 Observed piezometric pressures

the cutting process was thought to have induced a momentary increase in the pore water pressure in the voids of these sandy soils, resulting in some loss of bearing capacity. This magnitude of settlement however was considered to be insignificant and did not have any adverse effect on the building structure. It had been reported by Luo *et al* (1997) that ground disturbance caused by installation of sheetpiles was significantly greater than that caused by jet grouting activity using the triple tube technique.

Substructure Construction

The excavation was carried out about 1 month after completion of the last jet grout column. The excavation was carried out without shoring. Upon exposure of the excavation, the protrusions of the grouted wall into the basement area had to be carefully trimmed by hand using chisels to avoid breaking the grout. The face of the jet grouted wall and slab were dry in general, except for areas of overlaps between jet grout columns which had not been properly formed due to the natural stiffness of the ground. These localised areas of seepage however occurred as damp or wet patches and no free flowing water was observed. Upon excavation to the final formation level at 3.5m deep, a 250mm diameter drilled cast-in-place micropile was installed from this level using down-the-hole pneumatic hammer for dynamic load test. The test micropile was installed over 4 days from 18 Nov 94 to 22 Nov 94. The remaining micropiles were installed between 29 Dec 94 and 19 Jan 95. The structural slab and walls of the new basement were cast on 8 Feb 95 and 23 Feb 95 respectively to complete the substructure.

The excavation for the new basement caused a relative displacement of about 7mm in I1 towards the excavation since the completion of jet grouting works (Fig.6). Upon completion of the micropiles there was another 4mm displacement towards the excavation in I1. Fig.7 however shows a progressive shift of about 5mm to 6mm in I2 away from the excavation. This inconsistency could have been caused by other activities on site close to inclinometer I2. As indicated by the water standpipes (Fig.8) there was significant drawdown of the groundwater during the micropiling works which had to puncture the grout slab to reach the founding soil strata below. The largest drop was recorded in WSP1 from +100.71mOD to +100.02mOD. For WSP2 the maximum reduction in water table was from +101.51mOD to + 101.38mOD. Changes in piezometric pressures were not significant in the clayey sand layer (Fig.9). Variations were generally between 4.8m to 5.0m for P1 and 4.6m to 5.0m for P2. This means that the clayey sand layer was not sensitive to the micropiling activity. Settlement of the structural columns were maintained within 3mm. The overall summary of the monitoring data corresponding to the critical activities is given in Table 3.

Table 3 Summary of monitoring data

Date	Activity	Lateral Displacement		Piezometric Head		Ground Water Level	
		I1 (mm)	I2 (mm)	P1 (m)	P2 (m)	WSP1 (mOD)	WSP2 (mOD)
29 Jun 94	Initial reading	0	0	4.3	4.9	+101.42	+101.26
22 Jul 94	Before jet grouting trial	1.19	1.92	4.1	4.8	+101.17	+101.27
29 Jul 94	1 day after jet grouting trial	6.02	3.18	4.0	4.5	+101.42	+101.48
5 Sep 94	39 days after jet grouting trial and before jet grouting works	6.97	2.77	4.9	4.7	+101.12	+101.37
24 Oct 94	1 day after completion of jet grouting works	8.77	3.43	4.8	4.9	+101.27	+101.84
21 Jan 95	Just after completion of micropiling works	7.47	9.00	5.0	4.8	+101.72	+101.53
24 Feb 95	After completion of base slab and side walls	8.68	8.02	4.8	4.7	+101.05	+101.57

Conclusion

The project demonstrated that with proper sequence of installation and operating parameters, the triple-tube jet grouting technique can be successfully applied to form grout columns beneath an existing building without having adverse effects on its foundations and superstructure. In the case where sandy backfill soil is present at shallow depth, a small settlement of the structures of the order of 3mm could be expected due to temporary rise in pore water pressures generated during the execution of jet grouting works. The grout envelope around the excavation pit provided sufficient strength and stiffness to retain the ground without shoring. Adequate watertightness was ensured to enable the new foundation and basement works to be carried out in reasonably dry conditions.

References

Ho C E (1995). An Instrumented Jet Grouting Trial In Soft Marine Clay, ASCE Geotechnical Special Publication No. 57, Ed. M J Byle and R H Borden, pp 101 - 115.

Luo S Q, Khoo B T and Ho C E (1997). Performance of Triple Tube Jet Grouting For An Underpass Excavation, Proc. 3rd Asian Young Geotechnical Engineers Conference, Singapore, pp 181 - 188.

ABUTMENT STABILIZATION WITH JET GROUTING

Michael J. Byle, P.E.[1] and Tarek F. Haider, P.E.[2]

ABSTRACT

Investigations of the east abutment of the existing Charles Street Bridge over Amtrak rail lines in Providence, RI indicated the existing structure would be adversely affected by proposed rail improvements. As a part of Amtrak's Northeast Corridor High-Speed Rail Electrification project, it is necessary to lower the track adjacent to an existing stone masonry bridge abutment by approximately 0.4 m (1.3 ft), reducing the stability of the existing structure. The rail line currently carries high speed passenger trains between Boston and Washington, DC. The bridge carries a heavily traveled road and a multitude of shallow underground utility lines including electric, gas, water, sewer, and telephone. The stabilization design required a method that would, as a minimum, maintain the pre-existing level of stability and minimize disruption to rail and vehicular traffic and utilities. Several conventional remediation alternatives were evaluated including underpinning and tiebacks. Jet grouting was selected because it was comparable in cost to the other methods and minimized disruption to utilities, bridge traffic and rail traffic. The jet grouting was designed to increase the mass and thereby the stability of the existing gravity abutment. The jet grout forms a soilcrete mass created by overlapping grout columns interlocking with the existing stone masonry. The jet grouting method permits construction through small diameter holes drilled through the existing roadway between the utilities, permitting both rail and motorized vehicle traffic to continue during the construction. The existing abutment was instrumented to continuously monitor ground heave and abutment movement during the grouting. The effectiveness of the jet grouting was verified by performing laboratory strength and density tests on core samples taken from the grouted mass and from on-site construction monitoring. The stabilized abutment has factors of safety in overturning, sliding, and bearing capacity meeting the requirements of current AASHTO design standards.

[1]Project Manager, Geotechnical Section, Gannett Fleming, Inc., 650 Park Avenue, Suite 100, King of Prussia, PA 19406
[2]Project Engineer, Geotechnical Section, Gannett Fleming, Inc., 650 Park Avenue, Suite 100, King of Prussia, PA 19406

INTRODUCTION

Jet grouting is a relatively new process, having been developed since 1970 (Kauschinger, et al, 1992). The process has only been accepted in the United States in the last 10 years. The increased demands for infrastructure rehabilitation in urban centers has created a need for new and innovative applications of technology. Jet grouting has been used for transportation infrastructure at airports (Ichihasi et al. 1992), tunnel construction (Kauschinger et al 1992, Xanthakos et al. 1994), and rail lines (DePaoli et al. 1989). Other infrastructure applications include creating gravity walls (Miyasaka et al. 1992) and underpinning buildings (Drooff et al. 1995, Xanthakos et al. 1994).

Amtrak's rehabilitation of its Northeast Corridor from Washington, D.C. to Boston, Massachusetts is called the High-Speed Rail Electrification project. Major improvements to rail lines including addition of overhead catenary electric lines has required modifications and improvements to numerous bridges, retaining walls, and other structures. One such structure is the Charles Street Bridge located in Providence, Rhode Island, where the tracks must be lowered 0.4 m (1.3 ft) to increase under-clearance to install the new electrification system components. Figure 1 shows the project location. The track lowering adjacent to the east abutment (Track 2) will affect the external stability of the abutment due to removal of soil support from the track side. The rail line is heavily traveled by high speed passenger trains and the roadway above is an equally traveled two-lane urban thoroughfare. Several shallow underground utility lines including electric, gas, water, and sewer are also being carried by the bridge. The stability of the abutment was investigated for the proposed undercutting and remedial measures were undertaken to stabilize the structure. The construction of the undercutting to lower the track will require a temporary undercut of an additional 0.6 m (2 ft).

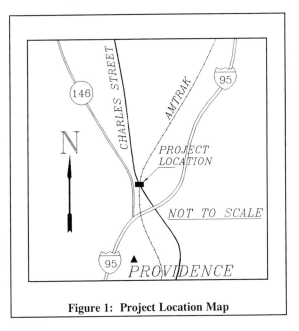

Figure 1: Project Location Map

STRUCTURE DESCRIPTION

The existing structure is a single span steel I-girder bridge supported by masonry abutments. The abutments are parallel to the tracks and the superstructure is at a 53⁰ skew. The existing horizontal clearance between the centerline of Track 2 and the face of north abutment ranges from 2.1 to 2.7 m (7 to 9 ft). The structure was partially reconstructed in 1977. The concrete bridge seats and backwalls were constructed over the original masonry abutments and the old superstructure was replaced. Existing drawings indicate that the east abutment is supported on shallow foundation bearing on soil. The north wingwall of the east abutment is parallel to Charles Street and will be unaffected by the grade lowering. The south wingwall gradually angles away from the tracks; therefore, the majority of the wall is far enough from Track 2 to be unaffected by the grade lowering. Consequently, the study was only limited to the east abutment of the bridge.

Visual inspection of the abutment revealed good condition of the individual masonry blocks and mortar joints. No voids were observed on the exposed face and no spalling was noticed on the top edges of the abutment. No soil erosion was noticed behind the south wingwall and the overall performance of the structure appeared adequate.

The bridge carries seven utility conduits, including four electric duct banks, one 406 mm (16 in) water main, one 304 mm (12 in) gas main, and one 508 mm (20 in) gas main. All of these utilities penetrate the backwall and extend over the abutment backfill. A 550 x 840 mm (22 x 33 in) brick sewer main crosses directly under the abutment, perpendicular to the abutment face, within 600 mm (18 in) of the bottom of the abutment.

SUBSURFACE CONDITIONS

The soil survey of Rhode Island (1981) maps the site under Urban Land and Udorthents - Urban Land Complex. These soils are predominantly loamy sand, sandy loam, fine sandy loam, or gravelly analogs. U.S. Geological Survey (1967) identified the site bedrock as relatively unmetamorphosed and generally well bedded sandstone and shale.

Two soil borings (B-1 and B-2) and four auger probes (P-1 through P-4), were drilled to determine the geometry and configuration of the abutment, and to provide information on the backfill and the abutment bearing medium. Locations of the borings and probes are shown on Figure 2. Boring B-1 was advanced through the masonry abutment into the supporting soil to determine its base elevation. Boring B-2 was drilled behind the abutment through the backfill to determine geotechnical properties of the backfill and the soil supporting the abutment. The auger probes were terminated at refusal on the back face of the abutment and hence, established the shape of the abutment. A cross-sectional sketch of the abutment approximated from the site exploration is shown on Figure 3.

The borings indicate the condition of the concrete and masonry of the abutment to be good, recording core recoveries from 91 to 100 percent in concrete and 72 to 100 percent in masonry. Granite masonry consists of hard and unweathered blocks, 0.5 m (1.5 ft) thick, with mortar joints between them. The bottom of the abutment was identified at a depth of 9.4 m (30.7 ft, Elevation 6.28 m). A split-spoon sampling of the soil beneath the abutment revealed an N-value of 100 blows for 100 mm (4 in) of penetration. The

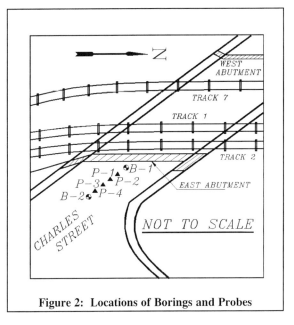

Figure 2: Locations of Borings and Probes

material sampled was coarse to fine sand with 1 to 10 percent silt. One inch of broken orange brick fragments was observed in the tip of the split-spoon sampler, which appears to have been the top of the brick sewer. The boring was terminated at 9.5 m (31.3 ft, Elevation 6.10 m).

The backfill behind the abutment was encountered in Boring B-2 to a depth of 13 m (40 ft). The upper portion of the fill, to a depth of 5 m (18 ft) is silty sand with gravel, with the Unified Soil Classification SM. The lower half of the fill consists of silt with sand (Unified Soil Classification ML). Based on the Standard Penetration Test N-values the overall density of the upper fill layer is very dense. The lower fill layer is generally in a loose to medium dense state.

The deeper natural soils penetrated by Boring B-2 consist of coarse to fine sand with 10 to 20 percent medium to fine gravel (SP-SM) with 1 to 10 percent silt. This layer extends from 13 m (40 ft) to the bottom of the boring at 15 m (49 ft). Standard penetration test N-values ranged from 18 to 53 and averaged 33 bpf. The 24-hour groundwater level was measured in Boring B-2 at 10.4 m (34.2 ft) depth (Elevation 4.85 m).

The auger probes, P-1 through P-4, were drilled to refusal as shown in Figure 3. The configuration of the back face of the abutment was interpreted as shown. The back face of the abutment is most likely stepped stone blocks rather than the smooth surface shown.

Soil strength parameters were estimated based on Standard Penetration Test N-values and laboratory tests, by utilizing existing correlations between N-values and soil parameters. The selected parameters are summarized in Table 1.

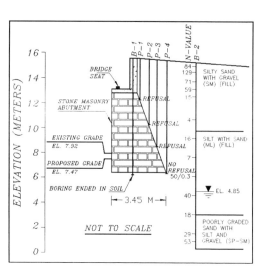

Figure 3: Cross-Section of Abutment Approximated from Site Exploration

STABILITY ANALYSES

Analyses were conducted to evaluate the stability of the abutment under current and proposed conditions. These analyses revealed that the abutment had very low factors of safety against overturning, sliding, and bearing capacity failure in the existing condition. The analyses also indicated that this would be aggravated by a proposed undercutting of approximately 0.4 m (1.3 ft).

Factors of safety were calculated for the existing abutment with the proposed undercut, as well as for the existing condition. Calculated factors of safety and corresponding RIDOT and AASHTO design standard values are presented in Table 2.

The calculated factors of safety for overturning, sliding, and bearing capacity were lower than AASHTO and RIDOT design standards for the existing condition. However, with the method of analysis used for this structure, the proposed construction will not reduce the factors of safety against sliding and overturning. Low factors of safety for overturning, sliding, and bearing capacity are expected when analyzing a gravity wall (or abutment) with low base width to height ratios. Acceptable base width to height ratios are typically on the order of 0.5 to 0.7. The pre-construction base width to height ratio for this abutment was approximately 0.3.

The calculated factors of safety against bearing capacity failure were less than 1.0. However, the structure was in good condition with no evidence of bearing capacity failure. Therefore, it was evident that current AASHTO method of analysis do not appropriately model the load distribution for this case. The likely explanation for the observed structure performance includes :

Soil Parameter	Abutment Backfill	Abutment Bearing
Angle of Internal Friction, ϕ	$34°$	$36°$
Moist Unit Wt., γ_m	115 pcf	120 pcf
Saturated Unit Wt. γ_{sat}	N/A	140 pcf

Table 1: Soil Properties

- Passive resistance from soil in front of the abutment not included in the analysis
- Redistribution of stresses during and/or after construction
- Strut action through the bridge superstructure

The proposed track lowering will reduce passive soil resistance and result in a lower factor of safety for bearing capacity which could result in failure. For this reason it was necessary to stabilize the abutment and bring it up to current design factors of safety.

	Overturning	Sliding	Static Global Stability	Seismic Global Stability	Bearing Capacity
RIDOT Design Standard	2.0	1.5	1.5	1.0	3.0
AASHTO Design Standard	2.0	1.5	1.5	1.1	3.0
Existing Condition	1.25	1.03	1.70	1.27	0.54
With Proposed Undercut	1.25	1.03	1.62	1.21	0.43

Table 2: Calculated Factors of Safety for Existing Abutment

FOUNDATION REMEDIATION ALTERNATES

A number of different alternatives were considered to stabilize the abutment. The stabilization criteria was based on a design method that would, as a minimum, maintain the pre-existing level of stability and minimize disruption to rail and vehicular traffic and utilities, at a reasonable cost. Because of the existing close clearance, encroachment into the existing limited track envelope was not an option. The following five possible remediation alternates were considered:

- Widening of footing by conventional underpinning
- Underpinning by jet grouting
- Replacement of backfill with light weight fill
- Installing tiebacks
- Enlarging the abutment by jet grouting through the backfill

A sketch of each alternative is shown in Figure 4. Table 3 presents the results of the evaluation including the advantages, disadvantages, and the procedure cost comparisons between alternates. In general, all cost estimates for remediation, including widening, underpinning with jet grout, replacing backfill with light weight fill, tiebacks, and enlarging the abutment by jet grouting backfill were based on final configuration factors of safety that meet AASHTO and RIDOT requirements. The lightweight fill option does not structurally change the existing abutment; and its evaluation was based on maintaining the current factors of safety during and after construction.

Comparative estimates of the stabilization costs fall within the range of $80,000 to $130,000. The two jet grouting alternates were both estimated to be the highest cost of available alternatives, as listed in Table 3. However, the duration of roadway disruption, cost of suspending rail traffic and risk of utility damage led to the selection of enlargement of the abutment by jet grouting of the backfill for remediation of this structure. This option results in the least impact to roadway and railroad operations and produces a stable structure that meets all of the AASHTO and RIDOT stability criteria. This option also minimizes interference with existing utilities, and does not require relocation or redesign of existing utilities. A significant consideration was that the strength required of the jet grouting behind the abutment is less critical than for underpinning or other bearing side improvements. This is because the grouted soil will act primarily as a dead weight, and need only have sufficient strength to transmit earth pressures to the existing structure.

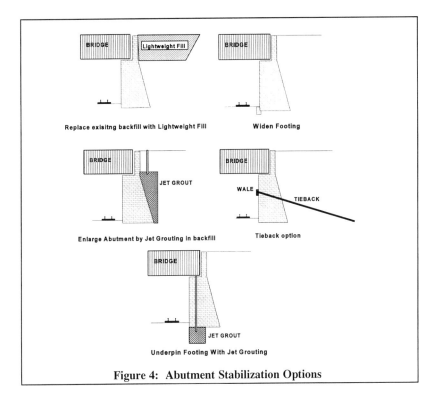

Figure 4: Abutment Stabilization Options

The lightweight fill option, though slightly less cost, will not give the same measure of stability as will the abutment enlargement option. The lightweight fill option would also require closing the bridge for an extended period and temporarily supporting a number of utilities including a 500 mm (20") gas main.

PROJECT CONSTRAINTS

The selected jet grouting scheme involved the creation of a soil-cement mass on the back face of the existing abutment. The following constraints were identified before execution of the project:

- Maintenance of single-lane bidirectional traffic at all times over the bridge.
- Restoration of full traffic flow at the end of each work shift.
- Drilling through and restoration of the reinforced concrete approach slab.
- Attaining adequate grout coverage in the zone of concern.

Alternative	Estimated Construction Cost	Estimated Construction Duration	Construction Considerations	
			Advantages	Disadvantages
Widen Footing by Conventional Underpinning	$80,000 + relocation of storm drain and fiber optic cable	5 weeks	• No impact to roadway • Relatively low cost • Increases stability	• Track side work requires track closure • Requires staged construction • Relocation of track drainage pipe required • Relocation of fiber optic line required • Footing extension will limit track options
Underpin with Jet Grouting	$130,000	3 weeks	• No track side construction • Maintains one lane of traffic during construction • No utility relocation required • Increases stability	• Short term street restrictions during construction • Protection of brick sewer under abutment needed • Grout acting in bearing must be verified at additional time and expense
Replace Backfill with Lightweight Fill	$100,000	3 weeks	• Maintains one lane of traffic during construction • No structural modifications required • No impact to rail traffic • Maintains existing stability	• Temporary sheeting needed for roadway traffic • Must provide temporary support for underground utilities during construction • Requires some restriction of roadway traffic
Tiebacks	$100,000	5 weeks	• No impact to roadway • No utility involvement • Increases stability	• Track side work requires track closure • Reduces track clearance
Enlarge Abutment by Jet Grouting Backfill	$130,000	2 weeks	• No impact to rail operations • Increases Stability • Short construction schedule • Maintains one lane of traffic during construction • Avoids impact to utilities	• Slightly higher cost • Requires some restriction of roadway traffic
No Action	$0	0	• None	• Unstable

Table 3: Evaluation of Alternates

- • Presence of nine shallow underground utilities.
- • Presence of a 550 mm x 825 mm brick sewer located below the foundation.
- • Control of abutment movement and ground heave.
- • Verification of the effectiveness of jet grouting.

Each of the above constraints was systematically evaluated and resolved during the design phase of the project. The constraints related to maintaining traffic flow during construction were tackled by preparing and executing a project specific traffic control plan.

DESIGN OF JET GROUTING

An iterative analysis was performed to optimize the shape of the jet grouted abutment and to ensure the required external stability. As shown on Figure 5, the jet grouting was designed to create a mass on the back face of the abutment which will shift the point of application of the resultant lateral earth pressure further into the backfill, increasing the abutment mass, and resulting in a higher resisting moment, larger effective bearing area, and reduced bearing stress at the toe of the abutment. Analyses indicated that jet grouting should extend from the back face of the abutment to 4.1 m (13.5 ft) behind the front face of the abutment. The required stability could be achieved with the top of the grouted mass beginning below the bottom of the existing utilities to the base of the abutment. The grouting was not permitted to extend below the existing bottom of foundation of the abutment to avoid potential interference with the existing brick sewer located only 0.45 m (1.5 ft) below the foundation bottom. Figure 5 shows a typical cross-section of the

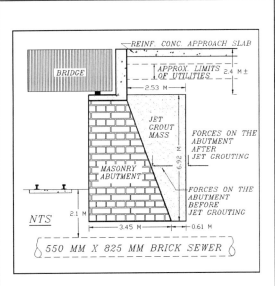

Figure 5: Cross-Section of Abutment Showing Jet Grout Mass

required jet grout mass. The estimated volume of the in-place stabilized mass was about 270 cubic meters (350 cubic yards).

The jet grouting was designed to provide sufficient overlap to produce a uniformly cemented mass to mechanically interlock with the back face of the existing masonry abutment. The jet grouting was designed in a staggered fashion to avoid excessive hydrostatic forces on the abutment from the liquid grout. Provisions were undertaken to maintaining a steady and constant grout waste return flow during the jetting process, since any imbalance in grout waste return would result in displacement of adjacent structures and ground heaving. A 350 mm (14-inch) thick reinforced concrete approach slab was located over the jet grouting zone which was underlain by the seven shallow underground utilities. Provisions were made to advance grout holes through the approach slab into the jet grouting zone, bypassing the underground utilities. Specifications required a minimum soilcrete density of 14 kN/m^3 (90 pcf) and a minimum unconfined compressive strength of 2,070 kPa (300 psi).

WASTE MANAGEMENT AND HANDLING

An erosion and sedimentation control plan was undertaken to prevent grout residue and other materials from the construction site entering the storm drains or other drainage systems or water courses. Perimeter sediment barriers, such as straw/hay bales and stone barriers were maintained at the site. All spoils and other waste products generated from the grouting operation were temporarily contained in an approximate 27 cubic meter (30 cubic yard) on-site waste pit. The spoils were allowed to set overnight and tested for contaminants. The uncontaminated wastewater was disposed of in the storm sewers and the solid portions of the waste were hauled to a dump site designated by Amtrak.

CONSTRUCTION OF JET GROUT COLUMNS

Jet grouting was performed using the triple fluid technique which pumped water at a low flow rate but at high pressure to break down the soil matrix and enhance soil erosion, while cement grout was pumped at a lower pressure to immediately fill the eroded spaces. The third fluid, the air jacket, protects the high pressure water jet and helps lift the grout spoils as it is exhausted up the drill annulus. Adjusting air and water jet pressures and flow rates during jet grouting allowed construction of grout columns in excess of 1.2 m (4 ft) in diameter. The grouting fluid pressures, flow rates, and monitor extraction speed were carefully selected for the type of soil to be jet grouted and the size of columns to be constructed, while avoiding ground heave or abutment movements resulting from excessive grout pressure or velocity. The water, air, and grout pressures during jetting were maintained around 40, 0.83, and 1.25 MPa (6,000, 180, and 120 psi), respectively. The grout flow rate was maintained at 150 litres/min under a monitor withdrawal rate of 30 cm/min.

The installation of the jet grout columns progressed as follows:

1. Pavement was saw cut behind the approach slab and all utilities were identified by hand excavation of the backfill. Utilities were marked on the grouting zone.
2. Strips of the approach slab were saw cut and removed to advance grout holes into the jetting zone, while keeping areas over the utilities covered by the slab. Roadway steel plates were used to maintain traffic through the construction zone.
3. Baselines were established for locating grout holes in areas where slab is removed. A minimum 450 mm (18-inch) clearance was maintained between the grout holes and all gas lines and 300 mm (1 ft) to other utility lines.
4. Grout holes were aligned according to the planned batter angles.
5. Grout holes were advanced through the soil fill to the jet grouting zone.
6. The required grout mass was constructed by overlapping individual grout columns by at least 1/8 of the column diameter or 150 mm (6 inches), whichever is greater.
7. Drilling and jet grouting were sequenced to maintain a minimum distance between freshly grouted columns. A minimum 24-hour curing period was required before installing adjacent columns.
8. A maximum 1.8 m (6 ft) spacing between individual grout columns was maintained.
9. The portion of the grout holes from the top of the grouting zone to the bottom of the approach slab were filled with grout without jetting.
10. After completion of grouting, the removed portions of the approach slab were repaired and restored with plain concrete and doweled to the original slab.

Figure 6 shows the approximate grout hole layout plan. Figure 7 shows a typical cross-sections of the abutment with individual grout columns and overlaps. Angling of the jet grout columns was included at the suggestion of the specialty grouting contractor, to maximize overlap while avoiding interference with the existing utilities. Some small wedges of soil are expected to remain in the grouted zone where utilities prevent clear access. However, these wedges will be relatively small inclusions in the upper third of the larger grout mass. These will be of no consequence since the lower grouted mass is contiguous.

MONITORING OF ABUTMENT MOVEMENT AND GROUND HEAVE\

Controlling movement of the abutment during construction was given immense importance because of masonry construction and current stability of the structure. The abutment was instrumented during the grouting to assess and arrest any grouting-induced movement. Abutment tilt was monitored with vertically placed "EL beam sensors" during the jet grouting. Five (5) beam sensors were mounted vertically on the abutment and connected to a remote automatic datalogger system. The datalogger system provided continuous real-time tilt readings of the abutment and temperature variation during jet grouting.

Since many factors affect tilting of the abutment other than jet grouting, the threshold value of abutment movement was established by monitoring movements caused by temperature variations, wind, and shocks induced by trains and vehicles. Prior to the start of jet grouting, a 48-hour continuous period of data in all beam sensors were recorded and the baseline readings for abutment movement were established at each monitoring location under normal operating conditions. The threshold for abutment movement was selected as the tilt angle recorded at any beam sensor which results in an additional 0.02 degree rotation over the established extremes provided that the movement does not cause damage or distress to existing utilities or structures.

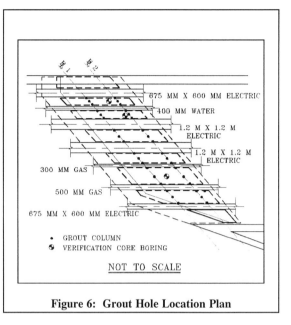

675 MM X 600 MM ELECTRIC
400 MM WATER
1.2 M X 1.2 M ELECTRIC
1.2 M X 1.2 M ELECTRIC
300 MM GAS
500 MM GAS
675 MM X 600 MM ELECTRIC

• GROUT COLUMN
◈ VERIFICATION CORE BORING

NOT TO SCALE

Figure 6: Grout Hole Location Plan

Ground heaving was monitored continuously and simultaneously at several locations on the pavement within the jet grouting zone. Heave monitoring was performed with a rotating laser level with targets set at selected points on the pavement behind the abutment. The targets were equipped with an alarm indicator that alerts the grouter of movements in excess of the established threshold. The threshold for heave of the pavement was considered 3 mm (1/8 in). During the course of the project, no abutment tilt or ground heave, above the threshold, was recorded.

VERIFICATION OF JET GROUTING

After construction of all jet grout columns, two continuous core borings of 127-mm (5-in) diameter were drilled through the grouted mass to verify the effectiveness of jet grouting. In addition, the contractor drilled a core boring, halfway through the jet grouting process, for his own quality control purpose. That boring indicated that the required size columns were being constructed by the jetting process. A total of six (6) core samples from the grout mass were tested for unit weight and unconfined compressive strength. The core borings satisfied the requirement of a minimum 70 percent core recovery. Laboratory test results revealed much higher density and unconfined compressive strength for the in-place, completed jet grout mass than the project requirements. Measured strength ranged from 10.4 to 25.2 MPa (1,510 to 3,660 psi) and density from 17.9 to 20 kN/m³ (114 to 127 pcf).

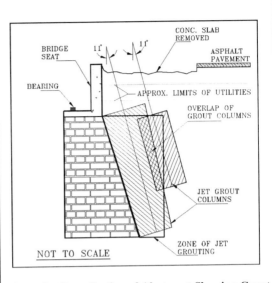

Figure 7: Cross-Section of Abutment Showing Grout Columns and Overlaps

CONCLUSIONS

Stabilization of the north abutment of Charles Street Bridge was achieved by jet grouting of backfill behind the abutment. The successfully executed jet grouting method of abutment stabilization provided a number of unique advantages . These include:

- least disruption to traffic flow over the busy street
- least impact to the underlying utilities
- no interruption of rail service
- verifiable
- achievable quality requirements
- economical
- short construction period

Recent applications of jet grouting have shown it to be a reliable system in the hands of a qualified contractor with the proper design and construction control. The application of jet grouting to augment existing structures should consider the structural properties and limitation of the method. The addition of a counterbalance grout mass to the back of an abutment is a viable application that requires material strengths well within the range of capability for the method. Jet grouting is a flexible and adaptable system that has great potential in infrastructure improvements.

REFERENCES

Kauschinger, J.L., Perry, E.B. and Hankour, R. (1992) "Jet Grouting: State-of-the-Practice", Geotechnical Soil Improvement and Geosynthetics. ASCE Geotechnical Special Publication No. 30. New Orleans, LA. February 25-28, p 169-181.

Ichihashi, Y., Shibazaki, M., Kubo, H., Iji, M. and Mori, A. (1992) "Jet Grouting in Airport Construction", Geotechnical Soil Improvement and Geosynthetics. ASCE Geotechnical Special Publication No. 30. New Orleans, LA. February 25-28, p 183-193.

Burke, G.K., Johnsen, L.F. and Heller, R.A. (1989) "Jet Grouting for Underpinning and Excavation Support", Proc. 1989 Foundation Engineering Congress, Evanston, IL, p 291-300.

Miyasaka, G., Sasaki, Y., Nagata, T., Shibazaki, M., Iji, M., and Yoda, M. (1992) "Jet Grouting for a Self-Standing Wall"Geotechnical Soil Improvement and Geosynthetics. ASCE Geotechnical Special Publication No. 30. New Orleans, LA. February 25-28, p 144-155.

DePaoli, B., Tornaghi, R., and Bruce, D. (1989) "Jet Grout Stabilization of a Peaty Layer to permit Construction of a Railway Embankment" Proceedings of ASCE Congress: Foundation Engineering: Current Principles and Practices, June 25-29, Evanston IL. p 272-290.

Drooff, E., Furth, A., and Scarborough, J. (1995) "Jet Grouting to Support Historic Buildings," Foundation Upgrading and Repair for Infrastructure Improvement. ASCE Geotechnical Special Publication No. 50. San Diego, CA, October 23-26, p 42-55.

Xanthakos, P., Abramson, L. and Bruce, D. (1994) "Ground Control and Improvement," John Wiley & Sons, New York, NY 1994 Chapter 8, "Jet Grouting" by Donald A. Bruce. p 580-683.

JET-GROUTED CANTILEVER WALL FOR SLOPE STABILITY

Allen J. Furth, Assoc. M.ASCE[1]
Steven A.Wendland, Assoc. M.ASCE[2]

ABSTRACT

A major electric transmission line structure located at the top of a 28 m high bluff was in danger of becoming unstable because of potential slope failure. Analysis indicated that the cliff would not reach a stable slope angle before undercutting the structure. Potential for disruption in the region's power grid prevented relocation of the tower away from the failing bluff. The solution to stabilize the existing foundations included lowering the ground surface in the vicinity of the structure to increase the stability of the cliff, installing soil anchors through the existing 3.35 m diameter drilled pier foundations to increase the lateral load carrying capacity of the piers, and constructing an in situ cantilever wall using jet grouting. During jet grouting, unexpected movement of the existing structure necessitated a remedial program of permeation grouting to strengthen the soils around the existing piers. This paper discusses design issues, selection, field construction, verification monitoring and testing related to the jet grouting of the in situ cantilever wall.

INTRODUCTION

A major electric transmission line structure is located at the top of a 28 m high river bluff. The steel, H-frame structure was constructed in 1984 and supports a three-circuit, electric transmission line crossing the Patuxent River. These lines provide most of the electricity for northern Calvert County, Maryland. Each of the two legs of the structure are supported by a 3.35 m dia., 12 m deep drilled pier foundation as well as guy wires that are anchored with two other drilled pier foundations. With the structure in danger of becoming unstable because of potential slope failure, a solution

---

[1]Project Manager, Hayward Baker Inc., 1875 Mayfield Road, Odenton, MD 21113
[2]Geotechnical Section Supervisor, Black & Veatch,11401 Lamar Avenue, Overland Park, KS 66211

to stabilize the structure's drilled pier foundations was developed. This solution included:

1. Lowering the ground surface 3 to 4 m at the top of the bluff,
2. Installing soil anchors through the existing 3.35 m diameter drilled pier foundations to increase the lateral load capacity of the piers, and
3. Constructing an in situ cantilever wall between the structure and the unstable slope using jet grouting.

The general arrangement of the work is shown in Figure 1.

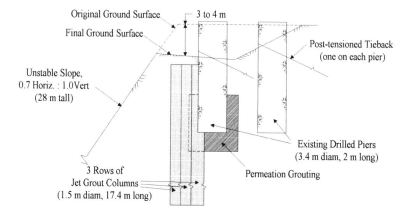

Figure 1. General Arrangement of Work

SITE HISTORY

The river bluff borders an undeveloped, pristine section of the Patuxent River in southern Maryland. The State of Maryland owns the land and has a strong interest in preserving the environment and the natural beauty of the river in this area.

The bluff has an approximate 0.7 horizontal to 1.0 vertical slope. When constructed, the transmission line structure was considered to be sufficiently located away from the top of the river bluff. However, the bluff is naturally unstable and has periodic shallow slope failures near the top. These shallow slope failures, common to the area, have caused the bluff to move closer to the structure.

In 1995, a shallow slope failure occurred in the bluff immediately adjacent to the structure, bringing the closest side of one of the drilled pier foundations to within approximately 2 m of the top of the bluff. Additionally, a tension crack with some vertical displacement was noticed intercepting the foundation closest to the cliff.

Moving the structure to a more stable location was not a viable option. If the structure was moved, two neighboring structures would need to be improved to compensate for increased spans and line angles between the structures. Also, moving the structure would have required an extended electrical outage to a large service area. The owner therefore retained a consultant to analyze the problem and develop approaches to stabilize the slope. Subsurface investigation was conducted early in the year for construction to begin by year end.

DESIGN ISSUES

Subsurface Investigations

The first phase of the design process involved performing a subsurface investigation at the site. Four soil borings were drilled at the site to collect samples for laboratory testing, perform standard penetration tests, and to determine the depth to the phreatic surface.

Site Stratigraphy

The borings indicated that the soil stratigraphy of the site was relatively uniform. The upper 20 m consisted of medium to loose sands and silty sands, below which was hard clay. The subsurface was modeled using a system of five soil layers. For each soil layer in the model, geotechnical parameters were developed for use in slope stability analyses.

Slope Stability Analyses

After the subsurface profile model was developed, a slope stability analysis was performed to determine the nature of the instability problem. Up to this point, slope failures at the site had been shallow. However, the analysis evaluated both potential shallow and deep failure surfaces to ensure that no stability problem was overlooked.

The stability analysis indicated that the slope was indeed unstable. The slope had a long term factor of safety of approximately 0.8 for shallow slope failures, indicating that more failures would continue until the slope had reached a shallower slope angle. The stability analysis indicated that potential deep failure surfaces, which would totally undercut the existing foundations, had a long term factor of safety of approximately 1.3. A factor of safety of 1.5 is considered to be the minimum for long term stability. However, since this river bluff had never exhibited any deep instability, all parties involved agreed that the stabilization efforts did not need to improve the stability of potential deep failure surfaces.

Stabilization Alternatives

Two categories of stabilization method were analyzed:

1. Methods that would prevent further slope failures.
2. Methods that would allow the structure to endure further slope movements without any loss of function.

The most frequently applied slope stabilization method to prevent further slope failure is the use of soil nails and/or tieback anchors working with a system of walers. However, the river at the base of the cliff and the irregular cliff face made access for this method very difficult. Additionally, environmental concerns, such as dumping of loose soils into the river and maintaining a natural appearance on the cliff face were of concern. Another way to stabilize the cliff would have been to use permeation grouting to increase the strength of a large volume of soil around the structure. However, large-scale permeation grouting would have been more expensive than the chosen option.

A second method considered was to make the slope angle shallower by placing fill at the toe of the slope or cutting the top of the slope. However, again environmental concerns prevented placement of a large amount of fill in the river.

The most logical alternative was to construct a retaining wall in the cliff. Such a wall would allow shallow movements to a degree, but would prevent de-stabilization of the structure by stopping the on-going movements at a certain point. The height of the cliff prevented use of a simple wall, such as cantilever sheet piles. Alternatives considered included walls constructed in-situ using auger-cast piles, micropiles, driven piles, or ground improvement methods. However, the presence of energized electric lines overhead, environmental concerns and the depth of the unstable soils limited the choices.

Ultimately, construction of an un-reinforced wall using jet grouting was selected. Jet grouting would allow construction of a wall at any depth or location and to any thickness or geometry. This option was approximately 30 percent less expensive than other acceptable options.

JET GROUTING TECHNOLOGY

Jet grouting was originally developed in Japan during the late 1960s and early 1970s, and has since found increasing acceptance worldwide (Bell et al., 1991). Within the past several years, there has been a significant increase in the technology's acceptance in North America. As jet grouting continues to evolve, it is proving itself in the emerging market of infrastructure improvement (Drooff et al., 1995) as well as in groundwater control and the more familiar application of structural support (Pellegrino and Bruce, 1996).

Jet grouting is often described as a high velocity erosion process (Kauschinger et al., 1992) that uses ultra-high pressures (300 to 600 bars) to impart energy to a fluid which is used as the soil cutting medium. This energy causes erosion of the ground and the simultaneous placement and mixing of the grout in the soil.

As shown in Figure 2, jet grouting has developed into three main systems, referred to as single-fluid, double-fluid and triple-fluid. Depending on the system being utilized, jet grouting can manufacture a product of cement slurry mixed with in situ soil, or nearly complete replacement of soil with an engineered cement slurry. The resulting final product is referred to as soilcrete (Furth and Deutsch, 1997).

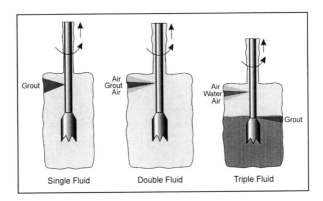

Figure 2. Jet Grouting Systems

The single-fluid system uses a high pressure grout jet(s) to erode the soil and mix it with the grout. In the double-fluid system, the addition of a second component of compressed air, shrouded around a high pressure grout jet(s) enhances the erosion and replacement effect of the high pressure grout jet(s). The compressed air helps to prevent the soil cuttings from falling back on the jet stream and helps to remove the soil debris by lifting it to the surface (Burke, Heller and Johnsen, 1989). In the triple-fluid system, three components (water, compressed air and grout) are injected simultaneously into the soil. The combined high pressure water jet, compressed air and grout volumes enables a higher percentage of soil to be removed from the ground and can be used for a nearly full replacement of the in situ soil with a cement grout product (Kauschinger, 1992).

DESIGN OF THE STABILIZATION SYSTEM

The depth and thickness of the wall required for stabilization were determined by modeling it as a cantilever wall, using two long term cases.

The first case determined the extent of continued slope movements that would be necessary to achieve a slope stability factor of safety of 1.0. After these continued failures, the upper portion of the soilcrete wall would be exposed and would act as a cantilever wall. The wall was designed to resist lateral earth pressures using a factor of safety of 1.5.

The second analysis case determined the extent of continued slope movements that would be necessary to achieve a slope stability factor of safety of 1.5. For this more stable shallower slope, a larger portion of the wall would be exposed. For this case, the wall was designed to resist lateral earth pressures using a factor of safety of at least 1.0.

For design, the wall was modeled as an unreinforced concrete member, designed with an unconfined compressive strength of 5.5 MPa. Probable difficulties in accurately placing steel reinforcement in the soilcrete precluded its use to strengthen the soilcrete. According to the analyses, the bending stresses controlled, and the wall's thickness was increased until it had sufficiently low bending stresses. The removal of soil around the existing drilled pier foundations would reduce their capacity to resist lateral loads, and uplift loads to a lesser extent. To restore the capacity of the existing piers, a post-tensioned grouted tieback was designed for each pier.

Selection of Soilcrete System

This project was originally put together in a competitive situation. Therefore the advantages to designing the installation for a large grouted mass wall hinged upon the effectiveness of creating larger diameter soilcrete columns. Based on the soil conditions represented in the borings, the proposed approach resulted in the selection of 1.5 m dia. soilcrete columns placed on 1.2 m center spacing and overlapping to create a continuous grouted mass. The double system was the preferred jet grouting technique for creating such large diameter columns since it was believed that this technique would be able to achieve both the geometry and the in situ soilcrete quality that the design parameters required.

The layout of the jet grout wall is shown in Figure 3. The wall was designed using three rows of overlapping columns, 1.5 m dia. and spaced at 1.2 m, installed approximately one meter below final grade down to a depth of approximately 18.6 m below final grade.

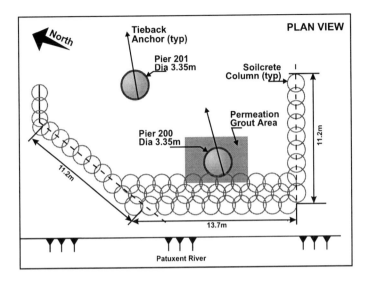

Figure 3. Jet Grouted Wall Layout

CONSTRUCTION OF STABILIZATION SYSTEM

Site Preparation

Prior to mobilization of the jet grouting equipment, a significant amount of preparation of the work area was required. To provide improved access for the jet grouting equipment under the energized high voltage lines and to further improve the stability of the slope, approximately 3 to 4 m of soil was removed at the top of the slope. The cut extended approximately 40 m along the bluff and 10 m back from the top of the bluff. Access into the area, which was located approximately half a mile off a paved road, required significant amount of road improvement in order for the trucks to maneuver. Additionally, measures were taken to prevent any contamination of wetlands located immediately adjacent to the work area. A temporary water well was constructed at the site to provide all the necessary water.

Jet Grouting Equipment

As a result of the large volumes of slurry which the double fluid jet grouting system needed to remain productive, the use of an automatic batching system was required. The slurry batching system consisted of a 50-ton silo with a 25-ton spare cement container placed next to an automatic colloidal 1 cubic meter mixer. The colloidal mixer was electrically controlled and created a slurry mix of water and cement with a specific gravity of 1.54. This batching system had the capacity of batching up to 30 cubic meters of grout per hour.

The jet grout pump was a triplex plunger pump with 100 mm pistons using a 530 HP engine with manual transmission. The drill rig consisted of a track mounted diesel hydraulic Casagrande drill with the jet grout controls based upon electronic control of the hydraulic system. A carousel was mounted to the drill in order to hold the full length of grouting rod and mechanically load the multiple sections necessary to install each jet grouting column to full depth.

Verification Testing

A quality control program was set up to ensure the quality of the grout mixing operations, the in situ geometry of the wall, and the in situ quality of the soilcrete. This included:

1. Routine density measurement of neat grout using a hydrometer.
2 Random sampling of the neat grout for unconfined compression testing.
3. Probing with the drill to determine the approximate in situ geometry of an individual soilcrete column (specifically the first two test columns) as well as to confirm the overall extension of the soilcrete wall.
4 Sampling of the newly created in situ soilcrete to confirm its in situ strength.
5. Retrieval of core samples from within the wall to determine consistency and to provide confirmation of the in situ sampling program.

Field Construction

The jet grouting mix consisted of a water/cement slurry with a specific gravity for the double rod system of 1.54. The cement was delivered to the site as a bulk material and stored in an upright silo. The cement was added via an auger into an automated mixer that weighed the water and cement components as they were mixed.

Before beginning production installation, a test program was performed which included the installation of two columns to full production depth. The next shift, the test columns were probed to determine an in situ geometry of 1.5 m diameter. Based upon the observed results of the probing, final production grouting for the double system was developed.

Once production grouting had begun, in situ soilcrete samples were retrieved using a drill-mounted, in situ sampler. As a minimum, one grouted column per shift was sampled and throughout the project 40% of the columns were sampled. These samples were cast into cylinder molds for unconfined compression testing.

Remediation for Structure Movements

Throughout the work, the position of the existing piers was surveyed daily to monitor any movements. Installation of the jet grout columns would temporarily cause a significant temporary reduction in soil strength. There was a concern that this disturbance would cause unacceptable horizontal foundation movements.

The maximum allowable horizontal movement for the foundations of the steel tube H-frame structure was 76 mm. Based on previous similar projects, such movements were not anticipated. However, early in the jet grouting work, the piers began to move measurably with each jet grout column installation. The probable cause was thought to be greater than expected disturbance of the soil. At once, it was obvious that the maximum allowable movement would be reached. A remediation program was immediately initiated in response to the unanticipated movement.

Earlier Installation of Tiebacks

Installation of post-tensioned tiebacks through each pier was originally planned to be installed after the completion of jet grouting work. To arrest the lateral movements of the piers, the jet grouting work was stopped, and the tiebacks were installed earlier than planned.

Permeation Grouting Program

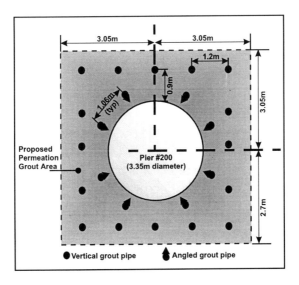

Figure 4. Permeation Grouting

Originally, the capacity of the pier closest to the bluff was to be increased by four jet grout columns installed around the pier. However, it was obvious that installation of these jet grout columns would cause significant movements. Therefore, a system of sodium silicate chemical permeation grouting was instituted to increase the strength of the soil around the pier significantly without disturbing the soil. This helped to limit pier movements when the jet grouting work continued. Approximately 200 cubic meters of soil around the bottom of the pier were grouted, as shown in Figure 4. Project costs increased by approximately ten percent because of this change.

Reassessment of Jet Grouting System

As the remediation program continued, evaluation of the in situ soilcrete samples indicated that the early strengths results for the double fluid jet grouting system were lower than anticipated (Figure 5).

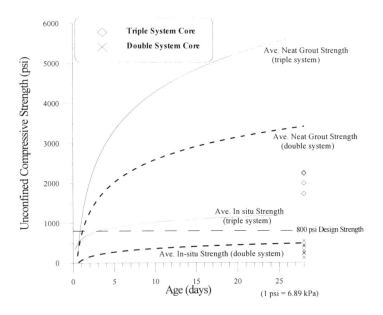

Figure 5. Unconfined Compressive Strength Results

Based upon the interpolation of these preliminary results, it was evident that the minimum 5.5 MPa strength requirement might not be achieved by this grouting method. This was attributed to a higher fines content than anticipated, and the presence of a thin, cemented sand layer at a depth of 13.5 meters.

Immediate plans were made to replace the double fluid system with a triple fluid jet grouting system. The triple system provides independent control of the cutting jet and injection pressures and volumes. It is also more efficient in the flushing of fines. Therefore, the change to the triple system would offer better protection against movement of the piers and, more importantly, provide the required in situ soilcrete strength. However, in order to maintain the original soilcrete column layout, the grouting parameters were changed to compensate for the difference of the two grouting systems. Principally, the grouting lift rate was significantly reduced to approximately half the rate in order to achieve the same design geometry as the double fluid system.

Cessation of Movements

The two changes discussed above and the switch to a triple fluid jet grouting system worked to decrease soil disturbance. These changes also increased the capacity of the pier, allowing it to better withstand temporary soil disturbance. When jet grouting work resumed, no more significant movements of the piers occurred.

Long Term Monitoring of Structure

The site will be subjected to periodic inspections to monitor any further slope failures, structure movements, and performance of the jet grouting work. This periodic inspection consists of annual site visits and surveys. The site visits will consist of looking for evidence of slope movements at the top and bottom of the bluff and measurement of the exact positions of the existing piers. If there is evidence of slope movements, topographic surveys will be performed if necessary to determine the magnitude and locations of slope movements. If the soilcrete wall is ever exposed by continued slope movements, the wall will be periodically inspected for deterioration and continuity.

CONCLUSIONS

The jet grouting work was completed in an environmentally sensitive area, with difficult access, no on-site water supply, and with energized high voltage lines overhead. The soilcrete cantilever wall constructed using jet grouting provided a predictable stabilization method that could be subjected to necessary quality assurance and quality control testing to ensure proper construction.

During jet grouting, unexpected movement of the existing structure necessitated a remedial program of chemical grouting. Lower than anticipated strength results for double-fluid jet grouting precipitated a change to the triple-fluid method. These design changes, plus early installation of tieback anchors, resulted in reduced soil disturbance and arrest of lateral movement. Despite the costs incurred in revising the program of work, jet grouting proved to be a versatile, cost-effective ground improvement technology for this site, reducing costs approximately 25 percent compared to other acceptable solutions.

REFERENCES

Bell, A.L., Crockford, R.M. and G.D. Mandley (1991). "Soilcrete Jet Grouting in Tunnel Construction in Cohesive Soils at Burnham-on-Sea, Somerset, England." *Proceedings, 6th International Symposium on Tunneling,* London, pp 249-261

Burke, G.K., Heller, R.A. and L.F. Johnsen (1989). "Jet Grouting for Underpinning: The Cutting Edge." *Geotechnical News, Volume 7, No.1*

Drooff, E.R., Furth, A.J. and J.A. Scarborough (1995). "Jet Grouting to Support Historic Buildings." *Proceedings of ASCE National Convention on Foundation Upgrading and Repair for Infrastructure Improvement.* San Diego, Ca. pp 42-55

Furth, A.J., Burke, G.K. and W.L. Deutsch (1997). "Use of Jet Grouting to Create a Low Permeability Horizontal Barrier below an Incinerator Ash Landfill." *Proceedings of International Containment Technology Conference.* St. Petersburg, Fl. pp 499-505

Kauschinger, J.L., Perry, E.B. and R. Hankour (1992). "Jet Grouting: State of the Practice." *Proceedings of Grouting, Soil Improvement and Geosynthetics.* New Orleans, LA. pp 169-181

Pellegrino, G and D.A. Bruce (1996). "Jet Grouting for the Solution of Tunneling Problems in Soft Clay." In *Grouting and Deep Mixing* (ed. Yonekura, Terashi and Shibazaki). Balkema, Rotterdam. pp 347-352

MICROPILE APPLICATION FOR SEISMIC RETROFIT PRESERVES HISTORIC STRUCTURE IN OLD SAN JUAN, PUERTO RICO

by

Brian H. Zelenko, P.E.[1], Donald A. Bruce, Ph.D., C. Eng[2],
David A. Schoenwolf[3], P.E., Robert P. Traylor[4]

ABSTRACT

As part of planned renovations and alterations to the existing U.S. Post Office and Courthouse (USPO&C) in Old San Juan, P.R., a seismic retrofit of the existing timber pile supported structure consisting of supplemental drilled micropiles was specified. This retrofit was required to bring the building into conformance with the 1987 upgrade of the Puerto Rico Building Code relative to seismic design. This paper presents a case study illustrating design details, construction procedures, and load testing data.

INTRODUCTION

The existing U.S. Post Office and Courthouse building in Old San Juan, Puerto Rico, was originally constructed in 1914. A major addition followed in 1940, and the structure represents an important historical landmark in the 500 year old harborfront area of the city (Figure l). The building is founded primarily on Raymond Step Taper piles and timber piles and is underlain by potentially liquefiable fine silty sand.

---

[1] Senior Engineer, Haley & Aldrich, Inc., Silver Spring, MD
[2] Principal, ECO Geosystems, Inc., Venetia, PA
[3] Vice President, Haley & Aldrich, Inc. Silver Spring, MD
[4] Geotechnical Division Manager, Structural Preservation Systems, Inc., Baltimore, MD

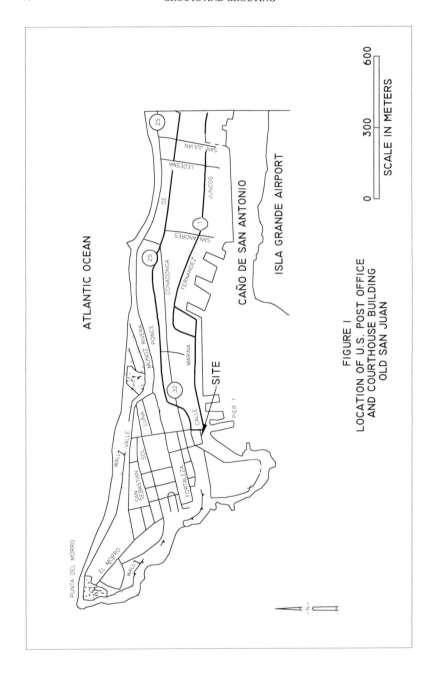

FIGURE I
LOCATION OF U.S. POST OFFICE
AND COURTHOUSE BUILDING
OLD SAN JUAN

As part of planned renovations and alterations to the structure, the owner, General Services Administration (GSA), specified a seismic retrofit including the installation of an enhanced foundation system. This retrofit was required, in part, to bring the building into conformance with the 1987 upgrade of the Puerto Rico Building Code, as it related to seismic design, considering the potential for a major seismic event on the island. Micropiles (FHWA, 1997) were selected for their constructability within an existing structure, ability to support the required loads, suitability for the subsurface conditions and cost effectiveness.

Analysis of the structure, the ground conditions and the potential seismic events led to the design of a system by the GSA and their consultants involving 217 micropiles, each with design service loads of 533 kN (60 tons) in compression, 356 kN (40 tons) in tension and 44 kN (5 tons) in lateral capacity, at a maximum allowable deflection of 13 mm (0.5 in.). This system of micropiles was designed to supplement the existing foundation system's ability to withstand anticipated loads associated with a design seismic event (90 percent probability of not being exceeded in 50 years).

GEOLOGICAL AND SITE CONDITIONS

Soil and Groundwater Conditions

A geotechnical investigation (Vazquez, 1994) revealed that the site is underlain by three strata (Figures 2 and 3):

Stratum I: The upper stratum consists primarily of miscellaneous man-placed fill and is approximately 2.4 to 3.0 m (8 to 10 ft.) thick. The fill is comprised of fine to medium grained sand with variable amounts of silt, clay and gravel. Standard Penetration Test (SPT) "N" values ranged from 3 to 100 blows per 0.30 m.

Stratum II: Below this upper fill and extending to a depth of 10.7 m (35 ft.) below ground surface, is a zone of fine to coarse grained sand. This stratum, which is typically 7.6 m (25 ft.) thick, was also found to contain organic silt and clay, cemented sand fragments, fine to coarse shell sand, coral fragments and peat. The relative density of this stratum ranges from very loose to very dense. SPT "N" values ranged from weight of hammer (WOH) to 49 blows per 0.30 m. Based on the low SPT "N" values in these saturated soils, the lower portion of Stratum II was judged by the owner's consultants to be susceptible to liquefaction in a design seismic event.

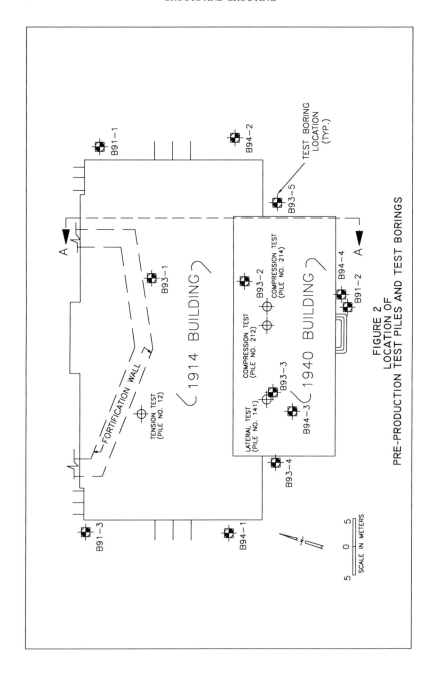

FIGURE 2
LOCATION OF
PRE-PRODUCTION TEST PILES AND TEST BORINGS

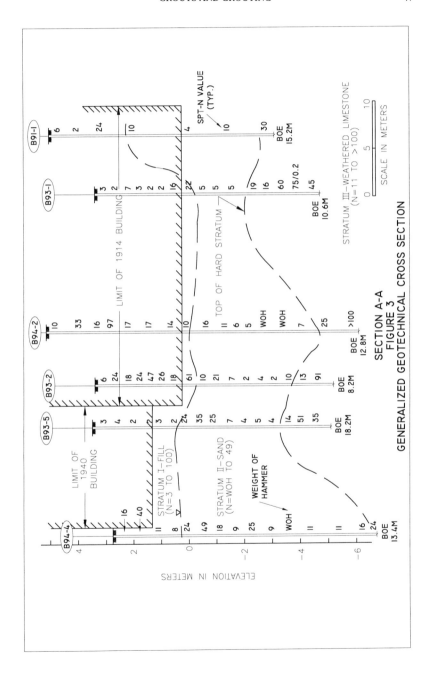

SECTION A-A
FIGURE 3
GENERALIZED GEOTECHNICAL CROSS SECTION

Stratum III: Below the sand stratum is a layer of weathered limestone and stiff clay. The stratum was described as medium to very hard, sandy to silty clay and clayey silt. This clay stratum was referred to in the contract documents as the "Hard Stratum" and was anticipated as the bearing stratum for the micropiles. SPT "N" values ranged from 11 to greater than 100 blows per 0.30 m.

Groundwater levels were anticipated to range in depth from 2.4 to 4.2 m (8 to 14 ft.) below existing ground surface level.

The ground surface elevation around the USPO&C varies between approximately El. 7.3 m (24 ft.) on the north side of the site to El. 2.6 m (8.5 ft.) on the south side.

Site Conditions

The structure is surrounded by very narrow streets which are frequented by many tourists. The building is six stories tall and is a massive masonry wall bearing structure with internal columns. The 1914 wing is founded partially on massive piers (northerly section) and partially on timber piles. The timber piles were driven to the top of Stratum III. The 1940 addition appears to have been founded on Raymond piles.

The micropiles required for the seismic upgrade were all located inside the building, where access was often severely limited, and work was conducted in low headroom conditions. During installation of the foundation piles, a number of site specific challenges were faced by the micropile contractor, including:

☐ Simultaneous overhead demolition and site clearance by another contractor;

☐ Encountering an old basalt block fortification wall dating from the 1600's;

☐ Existence of numerous other "obstructions" such as the existing 406 mm (16 in.) diameter timber piles, 1.8 m (6 ft.) thick pile caps, masonry footings and the remains of walls and footings from a demolished customs house building; and,

☐ Water rationing during a period of severe local drought and high ambient temperatures, which severely impacted drilling and grouting activities.

These challenges were each overcome by judicious initial design, and a flexible and responsive construction effort.

MICROPILE DESIGN

To date, micropiles have been infrequently used outside California when significant lateral loads, such as the 44 kN (5 ton) design capacity specified for this project, are anticipated. Conventional piles with a larger section modulus, such as steel H, concrete-filled steel pipe, or precast concrete piles have been selected. However, due to the special technical and logistical requirements of this project, micropiles proved the most appropriate choice. Special design considerations and the following criteria were stipulated by the Owner's consultant:

☐   The piles had to have a minimum outside diameter of 241 mm (9.5 in.);

☐   The pile design was to assume a "pinned connection" to the pile cap. Lateral pile load capacities and bending moments are affected by the type of connection. A pinned pile top connection (free to rotate) will have larger lateral deflections but the pile will experience smaller bending moments than a fixed (restricted against rotation) connection. Based on the Owner's consultant's analysis of existing piles at the site, they specified a pinned connection. The maximum allowable deflection of the pile subjected to the design lateral load of 44 kN (5 tons) at the top was 13 mm (0.5 in.);

☐   The micropile outer steel casing had to remain in-place and had to be socketed at least 0.9 m (3 ft.) into the "Hard Stratum" underlying Stratum II. However, the bottom of the steel casing had to be terminated no less than 7.6 m (25 ft.) below the top of the existing floor slab;

☐   The steel casing had to have a minimum wall thickness of 13 mm (0.5 in.); and, in the design calculations the wall thickness of the outer steel casing had to be reduced by 3 mm (1/8 in.) to allow for potential corrosion of the pile;

☐   The top 3.05 m (10 ft.) of the reinforcing steel in the micropile had to be encased by a smooth PVC sleeve sealed on both ends so as to preclude intrusion by the grout. (i.e. to develop a "free length").

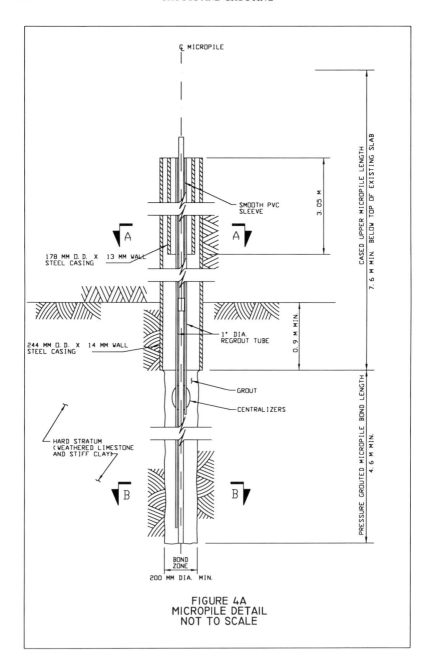

FIGURE 4A
MICROPILE DETAIL
NOT TO SCALE

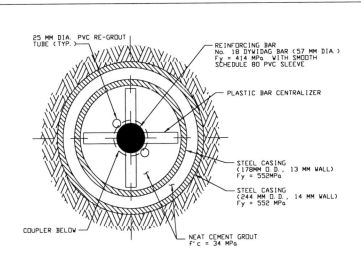

SECTION A-A
NOT TO SCALE

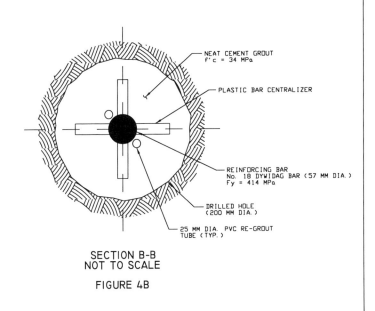

SECTION B-B
NOT TO SCALE

FIGURE 4B

☐   The 28-day compressive strength of the grout used in the piling could be no less than 34 MPa (5,000 psi); and,

☐   The maximum yield stress of the steel used in the micropile was limited to 552 MPa (80,000 psi).

As shown on Figures 4A and 4B, the approved design, prepared by Structural Preservation Systems and its subconsultants, incorporated a 244 mm (9-5/8 in.) diameter outer steel casing which was socketed a minimum 0.9 m (3 ft.) into the "Hard Stratum." Below the steel casing, an approximate 200 mm (8 in.) diameter hole was drilled a minimum of 4.6 m (15 ft.) into this stratum. A No. 18 steel Dywidag reinforcing bar was required for internal reinforcement. Additionally, a 3.1 m (10 ft.) length of 178 mm (7 in.) diameter, 13 mm (0.5 in.) wall thickness, 552 MPa (80,000 psi) yield stress steel casing was required in the upper 3.05 m of the micropile to provide sufficient resistance to lateral loads to satisfy the performance criteria.

While this second 178 mm (7 in.)steel casing is not typical for static micropile design, it was required to keep deflections below the allowable limit for the design lateral load. It was also required to keep the bending and shear stresses in the upper portion of the pile below the required limits.

INSTALLATION PROCEDURES

Micropile installation was performed with electro-hydraulic track-mounted drill rigs. The 244 mm (9-5/8 in.) diameter outer casing was drilled in 1.5 m (5 ft.) long sections, using water to flush the drill cuttings. This outer casing was drilled a minimum of 0.9 m (3 ft.) into the "Hard Stratum." A 4.6 m (15 ft.) deep bond zone was then drilled with a 200 mm (7-7/8 in.) diameter roller bit, again using water to flush the drill cuttings.

Once the hole was flushed clean of all drill cuttings, grout was pumped through the drill rods until fresh grout returned to the ground surface. The drill rods were then removed from the pile. The No. 18 Dywidag bar was then inserted into the pile with spacers and regrout tubes securely attached. The 3.05 m (10 ft.) long section of 178 mm (7 in.) diameter steel casing was then inserted into the top of the pile.

After all the steel reinforcing was installed, a pressure cap was screwed onto the top of the 244 mm (9-5/8 in.) outer steel casing, and the pile was pressurized to a maximum of 0.69 MPa (100 psi).

Grout was mixed in a colloidal mixer, and consisted of Type I/II with a water/cement ratio of 0.45. A retarder, Pozzolith 300R, was added at the rate of 0.12 x 10³ m³ (4 fluid oz.) per 0.44 kN (100 lbs.) of cement to ensure sufficient workability in the challenging site conditions, featuring high temperatures and long pumping distances.

PRE-PRODUCTION LOAD TESTING

Four load tests were conducted prior to production pile installation in accordance with applicable American Society of Testing and Materials (ASTM) standards to confirm the capacity of the installed piles when subjected to twice the design compressive, tensile and lateral loadings.

The contract specifications stipulated the following acceptance criteria for the various types of micropile load tests:

Compression Load Test

☐   Load vs. Movement curve is less than the theoretical elastic compression of the pile, plus 3.8 mm (0.15 in.), plus one percent of the pile diameter in inches;

☐   Net movement at the pile top is not greater than 13 mm (0.5 in.) following rebound from the maximum test load;

☐   150 percent of the design load, 801 kN (90 tons), must reach the top of the Hard Stratum during the maximum test load; and,

☐   The maximum test load is 1068 kN (120 tons).

Lateral Load Test

☐   Maximum allowable deflection at the top of the pile is 13 mm (0.5 in.) at design load during the last cycle; and,

☐   Design load is 44 kN (5 tons), the maximum test load is 88 kN (10 tons).

Tension Load Test

☐   Net upward movement does not exceed 13 mm (0.5 in.) at the pile
      head after removal of the maximum test load;

☐   No continuous upward movement (creep) without increase in load;
      and,

☐   Maximum test load is 712 kN (80 tons).

Compression Load Tests (Piles 214 and 212)

      Pile No. 214, was the first pile installed, and a combination of
difficult ground and construction problems resulted in an effective bond zone
only 3.4 to 3.7 m (11 to 12 ft.) long. In addition, the global pressure
grouting method subsequently used on all piles, (i.e. through the head) was
not efficiently applied on this pile. It was, therefore, unlikely that the
decompressed ground around the bond zone was effectively recompacted
during the installation process. Nevertheless, the decision was made to
continue with the test, and the pile was loaded to 200 percent of the design
load (i.e. 1068 kN (120 tons)) using a 300 ton hydraulic jack and adjacent
production piles as reaction piles. After unloading (Figure 5), the movement
at the top of the test pile was 52.2 mm (2.056 in.), which was in excess of the
specified acceptance criterion.

      The pile was then regrouted through the post grout tubes, which had
been installed in the test pile as a contingency measure. The pile was
subsequently re-tested. As shown in Figure 6, the net movement at the top of
the pile, following rebound from the maximum test load, 1068 kN (120 tons)
was 3.20 mm (0.126 in.) The results of the re-test met the acceptance criteria
and the pile was approved as a production pile. A comparison of the
permanent movements recorded before and after post grouting clearly
highlights the benefits of the post-grouting operation.

      A mechanical telltale was installed in the test pile down to the top of
the Hard Stratum to evaluate the amount of load reaching there during the
load test. Additionally, as a backup to the telltale, the internal reinforcing bar
in the pile was isolated from the grout, from the ground surface to the top of
the hard stratum. A PVC sheathing was installed over the reinforcing bar
this entire length. A coating of grease was applied to the sheathing prior to
installation to reduce the bond between the grout and the sheathing. Below
each reinforcing bar coupling in the pile, styrofoam blockouts were installed
to allow the reinforcing bar to move without being restricted by the grout.

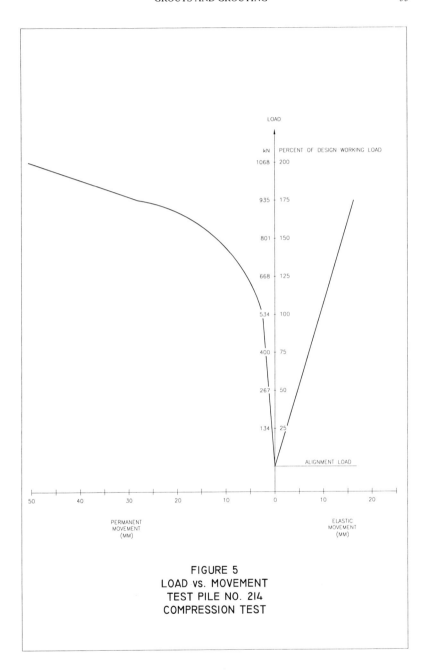

FIGURE 5
LOAD vs. MOVEMENT
TEST PILE NO. 214
COMPRESSION TEST

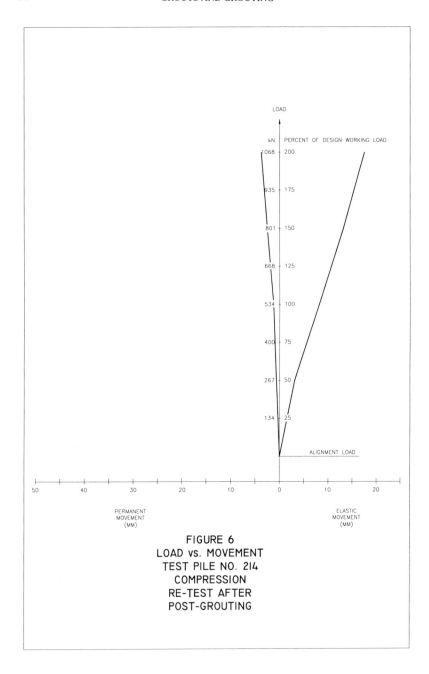

FIGURE 6
LOAD vs. MOVEMENT
TEST PILE NO. 214
COMPRESSION
RE-TEST AFTER
POST-GROUTING

Based on analysis of the telltale movements, the load at the top of the Hard Stratum was 827 kN (93 tons) when 1068 kN (120 tons) were applied at the pile top. This is equivalent to 155 percent of the design load and thus satisfied the specified criterion.

Nevertheless, given the uncertainties over the construction of Pile 214, a second compression load test was conducted on Pile No. 212. As shown in Figure 7, the measured total movement at the pile top was 15.2 mm (0.600 in.) at 1068 kN (120 tons), of which 5.41 mm (0.213 in.) was permanent. The analysis of the telltale movements indicated that the load at the top of the Hard Stratum was 667 kN (75 tons) when 1068 kN (120 tons) were applied at the pile head, equivalent to only 125 percent of the design load. Based on analysis of the elastic deflection of the reinforcing bar, it was determined that the free length was less than foreseen, and that the reinforcing bar must have bound up in the pile during testing. Nevertheless, acceptance of the pile was recommended since it was capable of supporting the design loads consistent with the intent of the designer. The compression load test was therefore approved.

Lateral Load Test (Pile No. 141)

The loading was applied with an Enerpac RC106 hydraulic jack with a rated capacity of 89 kN (10 tons). During the lateral load test, the lateral deflection was recorded as 11.4 mm (0.448 in.) at 44 kN (5 tons) design load on the last cycle during unloading from the 200 percent load. This was less than the 12.7 mm (0.5 in.) allowable deflection at the top of the pile and the lateral load test was approved. Figure 8 shows elastic and permanent movement recorded during the pile loading. Because the pile is laterally supported by soil, the elastic movement was non-linear.

Tension Load Test (Pile No. 12)

During the tension load test, the total movement at the pile top was recorded to be 19.3 mm (0.760 in.) at a maximum load of 712 kN (80 tons) with a permanent movement of 5.87 mm (0.231 in.) after unloading. The net upward movement did not exceed the specified maximum of 12.7 mm (0.5 in.) at the head of the pile after removal of the maximum test load. In addition, since no continuous upward movement (creep) without increase in load was observed, the tension load test was approved (Figure 9).

FINAL REMARKS

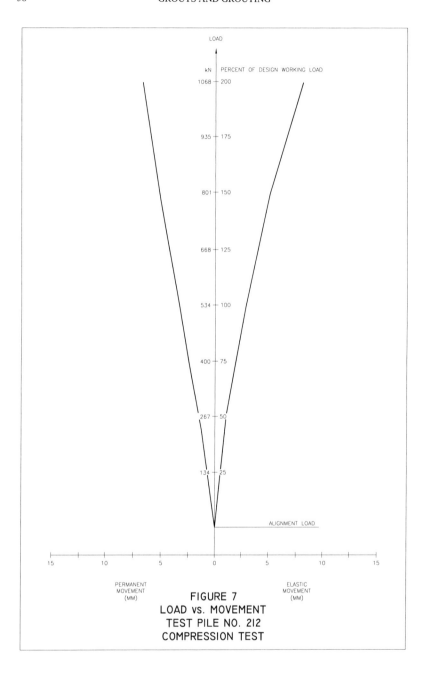

FIGURE 7
LOAD vs. MOVEMENT
TEST PILE NO. 212
COMPRESSION TEST

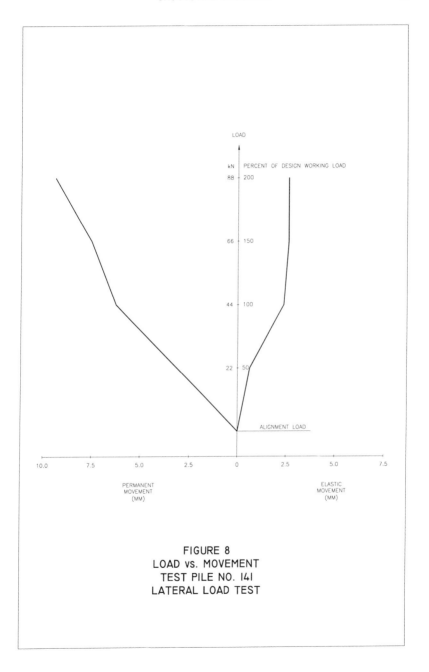

FIGURE 8
LOAD vs. MOVEMENT
TEST PILE NO. 141
LATERAL LOAD TEST

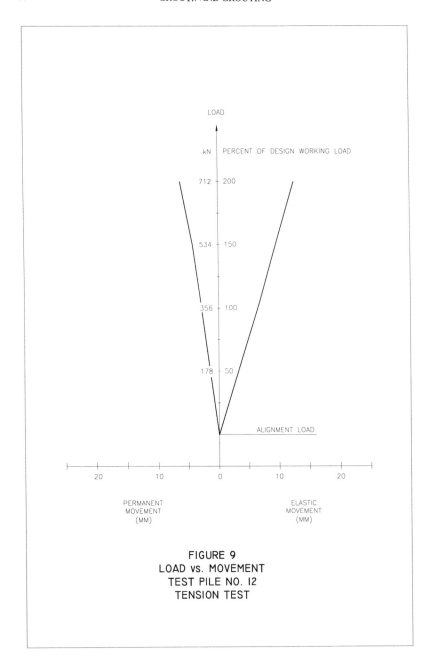

FIGURE 9
LOAD vs. MOVEMENT
TEST PILE NO. 12
TENSION TEST

Despite the challenging design requirements and the very difficult construction impediments, the micropile installation was successfully completed during a three month period in the summer of 1997. This case history again illustrates the adaptability of micropiles in such conditions, and further underlines their excellent performance in test loading programs in compression, tension and lateral loading.

As building codes are updated to address our better understanding of soil-structure interaction during seismic events, it is anticipated that more building renovations of historic structures, similar to the one described in this paper, will be required. Supplemental foundation support provided by micropiles has been utilized for scores of building renovations. However, slender micropiles typically have not been considered when there are significant lateral load capacity and deflection requirements, as is required for a seismic retrofit. Piles with a larger section modulus, such as steel H, concrete filled steel pipe, or precast concrete piles have typically been considered. As has been shown, properly designed and constructed micropiles can be utilized for seismic renovations. Micropiles may be used to help preserve other historic, architecturally significant structures in seismically active areas.

ACKNOWLEDGMENTS

The General Services Administration, New York, New York, was the Project Owner. The Project Architect was Finegold Alexander & Associates, Inc., Boston, Massachusetts. Metcalf & Eddy, Wakefield, Massachusetts, served as Project Structural Engineer. Vazquez Castillo, Vazquez Agrait & Associates, San Juan, Puerto Rico, served as Geotechnical Engineering Consultant to the Owner.

Trataros, New York, New York, was the General Contractor for this project. Structural Preservation Systems, Baltimore, Maryland, was the Micropile Subcontractor to the General Contractor.

ECO Geosystems, Venetia, Pennsylvania, provided Technical Support to Structural Preservation Systems. Haley & Aldrich, Inc., Silver Spring, Maryland, served as Geotechnical Consultant to Structural Preservation Systems, and designed the micropiles and load tests.

REFERENCES

Project reports, specifications, and drawings.

U.S. Department of Transportation, Federal Highway Administration. Drilled and Grouted Micropiles: State-Of-Practice Review. Vol. I-IV, Report No. FHWA-RD-96-016, 017, 018, 019.:McLean, Virginia. U.S. Department of Transportation, Federal Highway Administration, 1997.

ASTM D1143, "Standard Test Method for Piles Under Static Axial Compressive Load".

ASTM D3966, "Standard Test Method for Piles Under Lateral Loads".

ASTM D3689 "Standard Test Method for Individual Piles Under Static Axial Tensile Load".

SCOUR REMEDIATION BY PENETRATION GROUTING

By

Khaldoun Fahoum[1], Assoc. Member ASCE, and John A. Baker[2], Member ASCE

Abstract

During augercast pile installation for the Laboratory Building expansion at the Illinois American Water Company site in East St. Louis, Illinois, an exposed 46-cm (18-inch) water valve, located within the construction excavation, ruptured during the early morning hours of February 8, 1996. The resulting water stream created a scour in the surrounding sandy soil estimated to be approximately 7.5 m (25 ft) by 10.5 m (35 ft) in plan dimensions and 2.5 m (8 ft) deep below the bottom of the valve. The scour undermined the foundation of the adjacent reinforced concrete building, approximately 2.5 m (8 ft) away from the valve. The permanent foundation remediation consisted of filling the scour with 5-cm (2-inch) diameter clean crushed limestone, to approximately 0.6 m (2 ft) below the undermined foundation, and grouting the entire scour zone using penetration grouting techniques. Thirty one grout pipes were driven into the crushed rock to depths of 0.6 to 4.5 m (2 to 14 ft) and covering most of the exposed area. The grouting scheme resulted in the entire zone being stabilized with grout, which increased the strength of crushed rock and reduced its permeability. A new foundation was then constructed below the undermined footing leaving a 5-cm (2-inch) gap between the two foundations, which was then filled with drypack concrete. No structural problems have been reported since the completion of the grouting program on February 26, 1996.

Introduction

Foundations for the new Laboratory Building, at the Illinois American Water Company site in East St. Louis, Illinois, consisted of 59 augercast piles to be installed in the sandy alluvial deposit of the adjacent Mississippi River. The top of pile elevation was approximately 2.5 m (8 ft) below existing ground surface which

---

[1] Senior Engineer, Geotechnology, Inc., 2258 Grissom Dr., St. Louis, Missouri 63146.
[2] Principal, Geotechnology, Inc., 2258 Grissom Dr., St. Louis, Missouri 63146.
Tel: (314) 997-7440, email: kf@geotechnology.com

necessitated an excavation of equal depth across the entire building pad. As a result of this excavation, a 46-cm (18-inch) water pipe was exposed in the bottom of that excavation. In addition, the basement wall of the reinforced concrete Filter Building, immediately east of the Laboratory Building, was exposed within the excavation. The Filter Building basement walls were supported on strip foundations bearing on sand. During pile installation for the Laboratory Building, the exposed 46-cm (18-inch) water valve ruptured, possibly due to freezing temperature, during the early morning hours of February 8, 1996. The resulting water stream, driven by pressure reported to be as high as 600 kN/m^2 (90 psi), created a scour in the surrounding sandy soil. The scour was estimated to be approximately 7.5 m (25 ft) by 10.5 m (35 ft) in plan dimensions and 2.5 m (8 ft) deep below the bottom of the valve. The scour exposed augercast piles recently installed to support the Laboratory Building and undermined the foundation of the adjacent Filter Building, approximately 2.5 m (8 ft) away from the valve. The contractor immediately filled the scour with 5-cm (2-inch) diameter clean crushed limestone rock and temporarily shored the undermined foundation with a steel beam supported on two of the recently constructed pile caps, as shown in Figure 1.

FIG. 1 Photograph Showing the Crushed Rock-Filled Scour and theTemporary Shoring

        Underpinning the building foundation using a deep foundation system was not an option due to the presence of live piping adjacent to the building. Further, the pile caps were not designed to carry an additional load from the Filter Building that was imposed by the temporary shoring system. Therefore, the selected course of remediation consisted of grouting of the crushed rock to increase its strength, reduce its permeability, and reduce the potential for fines to migrate into the crushed stone, and by constructing a new foundation bearing on the treated clean rock to permanently support the Filter Building.

The augercast piles were designed as friction and end-bearing piles. Friction from the top 3 m (10 ft) of soils surrounding the piles were not incorporated in the design and ,hence, the scour did not affect the design strength of the piles.

Grouting Process

Background. Depending on the application and soil gradations and soil or rock conditions, several grouting techniques are used in practice, including penetration grouting, compaction grouting, chemical grouting, jet grouting, and others. Penetration grouting, sometimes called intrusion or permeation grouting, involves the low pressure injection of low viscosity fluid grout or mortar into gravel, porous soil or rock voids to reduce permeability and increase strength (Mitchell, 1970; Hunt, 1986; and Naudts, 1996). Effective compaction grouting does not depend on the grout entering into the soil, gravel or rock voids; a thick grout mass, under a relatively high pressure, is introduced into the soil mass to displace and compact the soil (Mitchell, 1970; Brown and Warner, 1973; Warner and Brown, 1974; and Welsh et. al., 1987). Chemical grouting is conducted by injecting a pure chemical solution into a soil mass to change its physical characteristics provide the soil mass with additional cohesion (Mitchell, 1970; and Sutton and McAlexander, 1987). Finally, jet grouting is defined by the ASCE Geotechnical Engineering Division Committee on Grouting (1980) as a "technique utilizing a special drill bit with horizontal and vertical high speed water jets to excavate alluvial soils and produce hard impervious columns by pumping grout through the horizontal nozzles that jets and mixes with foundation material as the drill bit is withdrawn". This procedure is relatively recent in the United States and can generally be used on soils other than alluvial soils (Munfakh et. al., 1987).

Grouting. Penetration grouting was considered the most appropriate method for this project due to the gradation of the clean rock and the underlying loose sand. The clean crushed rock, already in place, is uniformly graded with few fines, and the underlying sand consisted of uniformly graded fine particles which was in a loose state due to the disturbance resulting from water pressure.

The grouting was performed by Berkel and Company of Kansas City, Kansas and observed by an engineer from Geotechnology, Inc., of St. Louis, Missouri. Grouting was accomplished through a total of 31, 2-cm (3/4-inch) in diameter, as shown on Figure 2. Grout pipes were driven into the crushed limestone using a vibratory hammer.

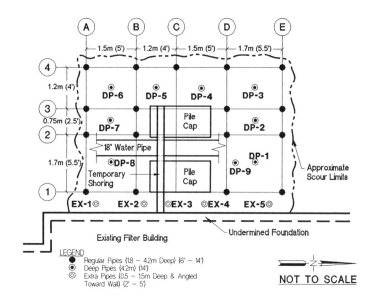

FIG. 2 Schematic Plan of Grout Pipes Layout

Shallow pipes (A-1 through E-4) were driven to a depth approximately 0.5 m (2 ft) below the crushed limestone layer corresponding to depths between 2 to 4 m (6 to 14 feet) below the surface using the grid system shown in Figure 2. The objective of using the shallow pipes was to grout the entire crushed limestone zone and as much as 0.5 m (2 ft) of disturbed soil below the limestone. In addition, nine deep pipes (DP-1 though –DP-9) were driven to approximately 4 m (14 ft) below the surface. Grout was to be pumped through the deep pipes after grouting through the shallow pipes was completed in order to further stabilize the zone of disturbed soil below the crushed limestone. Finally, five extra pipes were driven adjacent and parallel to the building foundation on an angle extending below the foundation to stabilize the zone immediately beneath the foundation. These pipes (EX-1 through –EX-5) were approximately 0.5 to 1.5 m (2 to 5 ft) in depth. Figure 3 is a photograph of the grout pipe layout.

Grouting commenced through the shallow pipes using a thin grout mix consisting of two bags of cement, one bag of fly ash and approximately 50 liters (13 gallons) of water. The purpose for using this mix was to assure that most of the voids are filled with grout and the interface zone between crushed rock and sand is sealed. Based on experience with other projects, such mix would readily fill the crushed rock and loose uniform sand voids, and would typically develop strength in excess of 14 MPa (2,000 psi). The grouted rock would provide a suitable subgrade for the strip footing which are typically proportioned for bearing pressures that are generally less than what an ungrouted crushed rock could provide. The ingredients were mixed on site with a double drum mixer, as shown in Figure 4. Approximately 3 m³ (110 ft³) of grout were pumped through the first pipe before moving to the other shallow pipes to pump approximately 0.3 m³ (12 ft³) in each of them. Several pipes were clogged and had to be pulled, cleaned, and redriven. Grout appeared at the surface around Pipes E-1 and E-2 while pumping in Pipes E-3, D-2 and D-3. At this stage the grout mix was changed to a thicker mix containing two bags of cement, one bag of fly ash, three bags of sand and approximately 64 liters (17 gallons) of water. This mix was used to provide additional bond between the crushed limestone particles. Such mix would typically develop compressive strength in excess of 20 MPa (3,000 psi). Using the thicker mix, approximately 0.6 m³ (20 ft³) of grout was pumped into each of the shallow pipes. This process was repeated, with the pipes being pulled up 0.5 to 1 m (2-3 ft) each time until the grout appeared at the ground surface. The grout level dropped slightly after pulling the pipes completely, but no additional grout was added in order to keep the grout level under the existing water pipe, as requested by Illinois American Water Company representative to maintain pipe access for potential future repairs. Grout was then pumped through the pipes adjacent to the foundation until the grout appeared at the surface. A tightening of the loose timber cribbing, placed immediately below the footing, was noticed while pumping through these pipes. The grouting pressure was approximately 350 kN/m² (50 psi) throughout the pumping operation, but was occasionally as high as 1000 kN/m² (150 psi). Such grout pressure is sufficient to fill the voids within the clean crushed rock.

FIG. 3 Field Layout of Grout Pipes

Finally, grouting was attempted through the deep pipes. However, at the maximum pressure of 2,400 kN/m^2 (350 psi) there was no grout flow. This phenomena was observed in Pipes DP-1 through -8. It was initially thought that the pipes were clogged, and an additional deep pipe, DP-9, was driven. This pipe, however, could not penetrate the previously grouted zone, and driving was terminated at 2.5 m (8 ft) below the surface. A 7 m (24 ft) long steel rod was then inserted into each of the deep pipes to clear any clogging and measure the inside cleared length of the pipes. All the pipes were cleared 12 to 18 cm (5 to 7 inches) beyond the length of the pipe except Pipe DP-2 which was clogged at approximately 3.8 m (12.5 ft) below the surface. Water was then pumped through the deep pipes to clear any remaining clogging but, again, at the maximum pressure the water did not flow through the pipes. At this stage additional pumping through the deep pipes was aborted, and the zone at 4.2 m (14 ft) below the surface was considered stabilized. The total grout volume pumped was approximately 29 m^3 (1050 ft^3) which corresponds reasonably to the void volume of the grouted zone.

FIG. 4 Grout Mixing Process

Foundation Underpinning. After the grout was set sufficiently, a new foundation was formed and cast to within 5 cm (2 inches) of the bottom of existing foundation. The 5 cm (2 inches) gap was ultimately filled with drypack concrete to complete the foundation underpinning.

Conclusions

The grouting process and foundation underpinning was a success where no structural problems have been reported since the completion of the project in February of 1996. The penetration grouting operation was completed within reasonable time frame (six days). The foundation remediation scheme described in this paper was an economical and effective alternative.

Acknowledgment

The authors would like to thank the Illinois American Water Company, in particular Mr. Kim Gardner, P.E., for granting permission to write this paper.

References

1. Brown, D.R., and Warner, J., (1973), "Compaction Grouting," Proc. ASCE, JSMFD, Vol. 99, No. SM8, pp. 837-847.

2. Hunt, R.E., (1986), Geotechnical Engineering Techniques and Practices, McGraw-Hill Book Company, New York

3. Mitchell, J.K., (1970), "In-Place Treatment of Foundation Soils," Proc. ASCE, JSMFD, Vol. 96, No. SM1, pp. 73-110.

4. Munfakh, G.A., Abramson, L.W., Barksdale, R.D., and Juran, I., (1987), "In-Situ Ground Reinforcement," In Soil Improvement-A Ten Year Update, ASCE Geotechnical Special Publication No.12, Edited by J.P. Welsh., pp. 43-55.

5. Naudts, A.A., (1996), "Grouting to Improve Foundation Soil" In Practical Foundation Engineering Handbook, McGraw-Hill Book Company, New York, Edited by Brown, R.W., pp. 5.277-5.400.

6. Sutton, J., and McAlexander, E., (1987), "Soil Improvement Committee-Admixture Report," In Soil Improvement-A Ten Year Update, ASCE Geotechnical Special Publication No.12, Edited by J.P. Welsh., pp. 121-135.

7. Warner, J., and Brown, D.R., (1974), "Planning and Performing Compaction Grouting," Proc. ASCE, JGED, Vol. 100, No. GT6, pp. 653-666.

8. Welsh, J.P., Anderson, R.D., Barksdale, R.P., Styapriya, C.K., Tumay, M.T., and Wahls, H.E., (1987), "Densification Subcommittee, Placement and Improvement of Soils," In Soil Improvement-A Ten Year Update, ASCE Geotechnical Special Publication No.12, Edited by J.P. Welsh., pp. 92-97.

LIMITED MOBILITY DISPLACEMENT GROUTING
FOR A MSE WALL FOUNDATION

Ching L. Kuo[1], Associate Member, ASCE
Wing Heung[2], Member, ASCE
John Roberts[3], Member, ASCE

ABSTRACT: During the construction of a Mechanically Stabilized Earth (MSE) wall in Section 3A of the Polk Parkway in Lakeland, Florida, excessive localized settlement was observed. Additional field explorations revealed the presence of a dump pit that included buried wood debris, tree trunks and possible voids below the wall foundation. A limited mobility displacement grout with a slump of 6 to 8 inches and grouting pressure of 200 psi was specified to fill the voids within the dump pit without running endlessly into surrounding voids or medium and to minimize the potential of overstressing existing wall panels. A monitoring program consisting of optical survey and tiltmeters was carried out as construction controls during the grouting injection operation to detect potential excessive movements on the existing wall.

INTRODUCTION

The Turnpike District of the Florida Department of Transportation has been constructing the Polk Parkway, a multi-lane limited access toll facility around Lakeland, Florida since 1996. The total length of the Parkway is about 24.5 miles, beginning at the Interchange of I-4 and Clark Road at southwestern side of Lakeland, and terminating at the interchange of I-4 and Mount Olive Road, northeast of Lakeland as shown in Figure 1. The project was divided into seven (7) design sections. Section 3 is about three (3) miles in length traversing in a west-east direction. The alignment of the Polk Parkway crosses many reclaimed phosphate mining areas. Aerial photographs dated 1941 and 1964 revealed that the alignment of Section 3 generally traversed heavily mined areas. These mined areas were subsequently reclaimed apparently commencing in the early 1980's. Small isolated undisturbed areas are also intermingled among these reclaimed areas.

---

1   Chief Engineer, PSI, Tampa, FL

2   Turnpike District Geotechnical Engineer, Parsons Brinckerhoff Quade & Douglas, Inc. Pompano Beach, FL

3   CEMC Program Director, Parsons Brinckerhoff Construction Services, Pompano, FL

Distress Observations

During the construction phase in 1997, a Mechanically Stabilized Earth (MSE) wall (Wall No. 1D) located west of the CSX railroad crossing experienced excessive localized settlement. A construction inspector initially noticed that the wall sagged over a span of five panel columns (Columns 25 to 29) as shown in Figure 2. A test boring (B-1) drilled in front of the middle of the sagged panels did not reveal any unusual soil conditions, which could have caused the excess settlement. It was decided to continue construction and the elevations of the concrete panels on top of the MSE wall were monitored periodically. During the first month after construction resumed, minor settlement of approximately 0.03 feet was observed over the entire length of the wall. As the construction progressed, the joints within the five sagged columns deteriorated as the panels rotated towards the center of the sagged section of the wall. As the rotation of these panels became more obvious, several panels near the top of the wall developed minor cracks. To determine the cause of the progressive distress, additional soil borings were performed to better define the subsurface soil conditions. At that time, the wall height was approximately 25 feet above the original ground surface and the remaining 5 feet of wall construction was stopped until the problem with differential settlement was resolved.

SOIL EXPLORATIONS AND GENERALIZED SUBSURFACE CONDITIONS

While performing the subsurface investigation during the design phase, soil borings were generally spaced at 100-foot intervals. Three soil borings, AB3-13, TB3-26 and TB3-27 were drilled during the initial subsurface investigation in the vicinity of the problem area of Wall No. 1D (PSI, 1994). The locations of these borings are shown in Figure 2. The soil conditions encountered consisted of a surficial layer of silty to clayey sands 5 to 10 feet thick underlain by sandy clay and silty sand strata ranging from 15 to 43 feet thick, see Figure 3. Based on these soil profiles, it appears that Wall No. 1D was founded on an undisturbed, unmined area.

After the wall panels between columns 25 and 29 began to show additional distress, a field testing program which included three new soil borings, B-2, B-3 and B-4 was performed to better determine the subsurface soil conditions in the problem area and to develop remediation alternatives. Figure 2 shows the locations of borings B-1 through B-4 and Figure 3 shows the soil profiles encountered at these locations. Borings B-1 and B-2 encountered soil conditions similar to those encountered during the design phase. However, boring B-3 revealed organic material and tree debris at a depth of 25 to 36 feet below the top of the MSE wall backfill and boring B-4 revealed the same debris at a depth of 25 to 28 feet. The SPT-N values (Standard Penetration Testing) ranged from 4 to 19 within the tree debris zone. However the SPT-N values do not reflect the true subsurface conditions because of the presence of the tree debris. The complete loss of drilling fluid circulation at several locations during the drilling operations also indicated the possible presence of voids and very loose conditions within the debris zone.

In addition to the new soil borings, the leveling pad between Columns 26 and 28 was exposed for inspection to help determine the cause and extent of the settlement problem. The survey data indicated that the leveling pad had settled approximately 5 inches after initial placement but surprisingly, the concrete pad did not show noticeable cracks.

Based on the soil boring information and the observation of the wall panel distress, it was concluded that a dump pit filled with tree debris and very loose soils existed under Wall No. 1D between columns 24 and 30. This dump pit was estimated less than 40 feet wide and 15 feet deep with 3 to 5 feet of soil cover atop the tree debris prior to construction of MSE walls.

SELECTION OF SOIL IMPROVEMENT METHODS

Remediation alternatives to stabilize the wall foundation were evaluated. In considering the working schedule, concerns of construction vibration, available working space, construction cost and the fact that Wall 1D was within five feet of final grade, deep soil improvement by grouting appeared to be the most feasible alternative to stabilize the foundation soils.

Four (4) basic types of grout are widely used in United States; slurry, chemical, jet grouting and compaction grouting (Elias, et al, 1996). Slurry and chemical grouting would fill voids between the soil particles with either cement or chemical binders. However, there was concern that due to the limited number of borings, the size and location of the dump pit was not accurately defined and if either of these two grouts were used, the amount of grout required might be excessive.

Jet grouting excavates soils by pressurized water jet and replaces with a mixture of excavated soils and cement grout. This would risk damaging the wall and the cost was estimated to be about two to three times greater than the compaction grouting alternative.

Compaction grouting compacts loose soils by injecting very stiff grout, usually 2 inch slump or less and at a relatively high pressure of up to 800 psi. However, the conditions encountered did not require compaction of loose soils and the efficiency of this thick, cohesive grout to fill the voids around the tree debris was uncertain.

For this specific site, the extent of the voids in the dump pit was unknown and the soils around the tree debris were very loose. The ideal grouting technique would fill the voids without using excessive grout and would compact the loose soils simultaneously. Unfortunately, this type of grouting technique is not possible with current technology.

After evaluating the different grouting methods, it was determined that filling the voids around the tree debris without using excessive quantities of grout was the essential requirement. Since none of the four common grouting methods described above was appropriate under these specific site conditions, alternative grouting techniques were investigated.

Limited mobility displacement (LMD) grouting was chosen as the preferred grouting technique. LMD uses a grout with a relatively high slump and low injection pressures. LMD grouting does not compact the loose soil but has been successfully used to fill voids (Byle, 1997). A slump of 6 to 8 inches and a relatively low injection pressure of 200 psi were specified for this grouting operation. It was expected that LMD grouting would provide sufficient flowability to fill the voids and the low grouting pressure would minimize grout loss.

LMD GROUTING PROGRAM AND RESULTS

A total of 24 grout injection holes were used in the grouting program as shown on Figure 2. The grout holes in the MSE wall embankment were located immediately behind the joints between the wall panels to avoid conflicts with the existing reinforcement meshes. Four (4) tiltmeters were installed on the face of the wall panels to monitor any deflections during the grouting operation. Ground heave was also measured by taking surveys shots of marks on a steel stand placed adjacent the grout holes.

The refusal criteria of the injection operation were specified as follows.

1. The injection pressure reaches 200 psi.
2. Ground heave of 1/4 inch or higher.
3. A grout injection quantity of 20 ft³ per one foot depth interval.
4. As directed by the Engineer according to deflections indicated from the tiltmeter instrumentation.

The grouting contractor started the grouting operation on January 15, 1997 and completed on January 28, 1997. At each grout hole, a 3-inch diameter grout pipe was installed first. In general, the grouting started at Elevation 130 feet (NGVD) and grout was injected at one foot intervals up to Elevation 140 feet (NGVD). At each injection point, one of the refusal criteria described above was reached before the grout pipe was lifted another foot or grout injection was terminated.

The grout mix design consisted of Type I or II cement, fly ash, silty sand and potable water. The design water cement ratio was 0.48. Samples of grout were tested for slump during the pumping process. The results of the slump tests ranged from 6 to 7.5 inches. Also, grout cubes were formed for compression testing. The compressive strength of the grout exceeded 600 psi after 24 hours setting. The injection pressures ranged from 50 psi to 350 psi. At some locations, the injection pressure increased very rapidly such that the pressure exceeded the 200 psi refusal criteria before the grout pump was stopped. The grout injection volumes ranged from 0.6 to 21 ft³ per foot interval. A total of about 1,370 ft³ of grout was injected in this project.

The program started with injection at Holes 1 through 4 in front of the MSE wall. It was expected that these injection holes would provide a grouted barrier, such that the subsequent grout injection would be confined below the wall foundation. Grout volume used in these holes was relatively high as shown on Figure 4. Some grout was detected to have traveled through soil medium and exited on the slope of a nearby retention pond.

LMD grouting behind the MSE wall included 10 primary injection holes, Holes 5 through 14, and 10 secondary holes, Holes 15 through 24. Figures 5 to 7 shows the grout injection volume versus the elevation at each injection hole. During the installation of grout pipes, grout was encountered at Holes 6, 7 and 8 from the previous grout injections at Holes 1 through 4 indicating that the grout flowed around the tree debris as expected.

As shown in Figures 4 to 7, only Holes 1 through 4, 6, 7, 10 and 14 reached the maximum volume at various depths. The majority of the secondary injection holes took only very limited amount of grout. This indicated that voids within the dump pit were filled mostly by injection through the primary injection holes. Injection Hole 5 was temporarily suspended at the beginning of the operation due to the presence of an existing storm sewer pipe. Later it was revealed that the adjacent Holes 9 and 15 accepted only very small grout volumes, which indicated that Hole 5 was probably outside the limit of the dump pit. On this basis, Hole 5 was removed from the grouting program.

The monitoring of the tiltmeters during the grout injection operation indicated that wall panels did not experience any significant movement. Limited ground heave was observed during injection at some locations. When ground heave exceeded 1/4 inch, grout injection at that hole was terminated. No major distress was noticed on the wall panels as a result of the ground heave.

CONCLUSIONS

After the LMD grouting program was completed, the cracked wall panels located near the top of the MSE wall were removed and replaced before the construction of Wall No. 1D and backfilling were resumed. The remaining 5 feet of fills and wall panels were placed to the final grade. Post-grouting soil testing was not performed because the presence of the tree debris was expected to affect the reliability of in-situ test results. However, the elevation survey performed on panels near the bottom of the wall indicated that no additional settlement was observed 2 months after the wall construction was completed.

Limited Mobility Displacement grouting with a slump of 6 to 8 inches was used on the project to improve the ability of grout to fill the voids around the tree debris. The maximum injection pressure was also limited to 200 psi to prevent undesirable movement of the existing MSE wall. The LMD grouting program has performed satisfactorily to fill the voids and reduce the post-construction settlement.

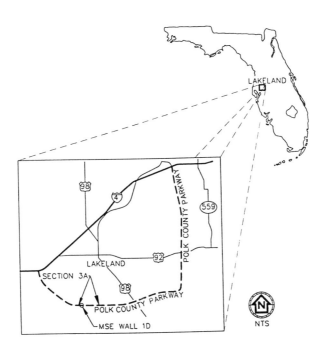

FIGURE 1 SITE VICINITY MAP

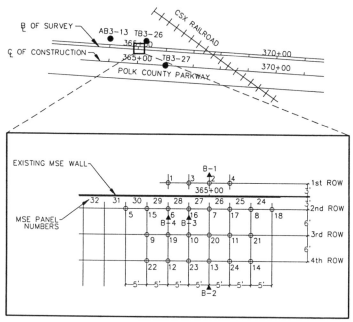

DETAIL A
GROUT INJECTION LAYOUT PLAN

PLAN SCALE

●   SOIL BORINGS PERFORMED IN DESIGN PHASE (1994)
▲   SOIL BORINGS PERFORMED IN CONSTRUCTION PHASE (1997)
⊕   GROUT INJECTION HOLE

FIGURE 2 SOIL BORING AND GROUT HOLE LOCATION PLAN

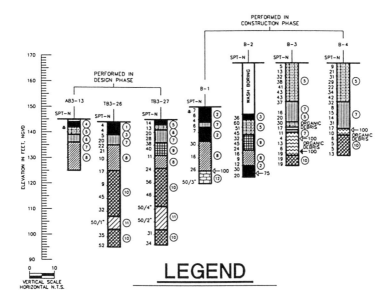

LEGEND

1. MIXED GRAYISH–BROWN SAND, TRACE SILT AND LIGHT GRAY CLAYEY FINE SAND

2. LIGHT GRAY TO BROWN SAND, TRACE SILT, SHELLS AND CEMENTED SANDS

3. LIGHT GRAY TO GREENISH–GRAY, OCCASIONAL ORANGISH–BROWN SAND, SOME SILT TO SILTY FINE SAND, TRACE PHOSPHATES, CEMENTED SANDS AND SILT

4. DARK GRAYISH–BROWN SAND, TRACE SILT AND ROOTS (TOP SOIL)

5. LIGHT BROWN TO GREENISH–GRAY SAND TRACE SILT, PHOSPHATES, SHELL AND ROOTS (TOPSOIL)

6. REDDISH–BROWN TO DARK REDDISH–BROWN SANDS, TRACE TO SOME SILT AND CEMENTED SANDS

7. LIGHT BROWN TO GREENISH–GRAY SAND, SOME CLAY TO CLAYEY FINE SAND, TRACE PHOSPHATES AND SHELL

8. LIGHT GRAY TO GREENISH–GRAY SANDY CLAY TO CLAY, TRACE PEBBLES AND CEMENTED SANDS

9. LIGHT BROWN TO GREENISH–GRAY SAND, SOME SILT TO SILTY FINE SAND, TRACE PHOSPHATES, LIMEROCK AND CEMENTED SANDS

10. LIGHT GREENISH–GRAY TO ORANGISH–BROWN SILTY FINE SAND TO SILT, TRACE PHOSPHATES AND CEMENTED SANDS

11. LIGHT GRAY TO ORANGISH–BROWN INDURATED CLAY/SILT, TRACE PHOSPHATES

12. LIGHT GRAY LIMEROCK, TRACE PHOSPHATES

FIGURE 3   SOIL BORING PROFILES

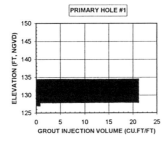

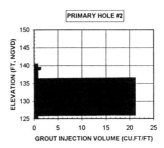

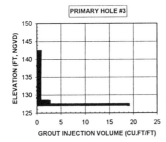

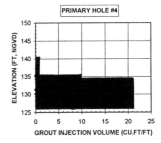

FIGURE 4   GROUT INJECTION VOLUME OF FIRST ROW

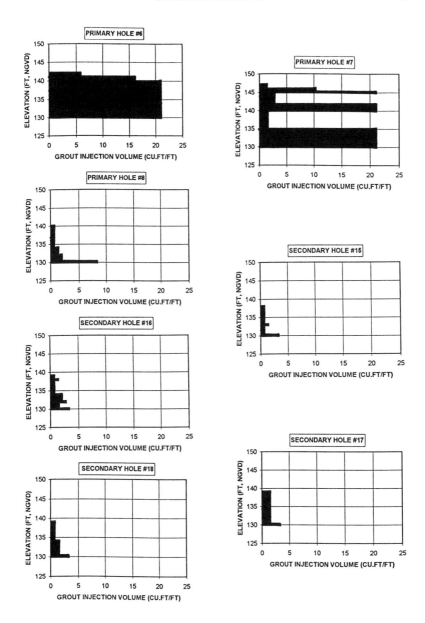

FIGURE 5   GROUT INJECTION VOLUME OF SECOND ROW

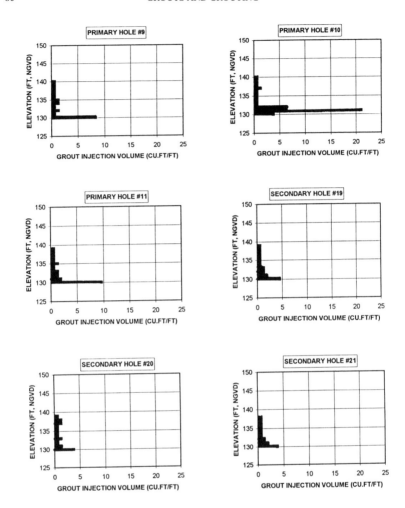

FIGURE 6   GROUT INJECTION VOLUME OF THIRD ROW

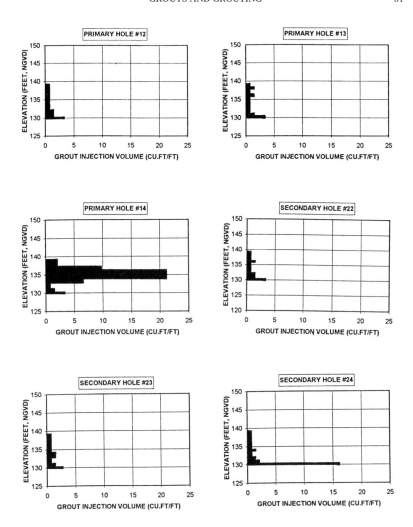

FIGURE 7   GROUT INJECTION VOLUME OF FOURTH ROW

ACKNOWLEDGMENT

Thanks are given to Jim Moulton Jr. of the Florida Department of Transportation, Turnpike District; Bill Adams of Parsons Brinckerhoff Construction Services, and Joe Chao of Kisinger, Campo and Associates for their dedicated efforts and excellent construction supervision and management. The grouting was performed by Hayward Baker, Inc.

REFERENCE

1.   Byle, M.J. (1997). "Limited Mobility Displacement Grouting: When Compaction Grout is Not Compaction Grout." ASCE Geotechnical Special Publication No. 66.

2.   Elias, V., Welsh, J., Warren, J. and Lukas, R. (1996). "Grout Improvement Technology Manual." FHWA Publication No. FHWA-Dp-3, Office of Technology Application, FHWA, Washington, D.C.

3.   Professional Service Industries, Inc. (1994). " Roadway Soil Survey - Polk County Parkway Section 3." Submitted to Reynolds, Smith & Hills, Inc./Florida Department of Transportation, Turnpike District.

REMEDIAL GROUTING OF DWORSHAK DAM

By W. Glenn Smoak[1], Member, ASCE, Francis B. Gularte[2], Member, ASCE
and Jerrold R. Culp[3], Member, ASCE

Abstract

Construction of the Corps of Engineers' Dworshak Dam, located on the North Fork of the Clearwater River near Orofino, Idaho was completed in 1972. At a structural height of 218 meters (717 feet), and an overall length of 1,002 meters (3,287 feet) at the crest, this dam is the highest, straight axis concrete gravity dam in the Western Hemisphere and the third highest dam in the United States. The dam was constructed on a foundation of granite gneiss that was slightly jointed and fractured having a permeability that decreased from about 1.1×10^{-3} cm/s (2.2×10^{-3} ft/min) at a depth of 9 meters (30 feet) to 5.1×10^{-7} cm/s (1×10^{-6} ft/min) at a depth of 76 meters (250 feet).

During the time period 1972 to 1984, drainage from all sources in monoliths 15 through 18 of the left abutment showed a relatively steady increase reaching a total of about 2,271 liters/min (600 gpm) at full reservoir elevation in 1984. Beginning in 1984, the rate of seepage from the gallery relief drains of these monoliths increased dramatically, reaching a total of 9,464 to 11,356 liters/min (2,500 to 3,00 gpm) in December of 1996. In the fall of 1995, a grouting contract was awarded to mitigate the increasing seepage. Cementitious grout was used to start the remedial grouting but abnormally high levels of regional precipitation quickly raised the reservoir elevation, and caused the Corps to terminate the work for safety reasons. This effort demonstrated, however, the difficulty that would be encountered in trying to grout the leakage with conventional cementitious grouts.

---

[1]Principal, Eco Structural Systems, LLC., 10438 W. Exposition Avenue, Lakewood, CO 80226

[2]Vice President, Hayward Baker Inc., 1780 Lemonwood Drive, Santa Paula, CA 93060

[3]Project Manager, Hayward Baker Inc., 140 W 2100 S, Suite 200, Salt Lake City, UT 84115

Figure 1. Photograph of the downstream face of Dworshak Dam. The zone of leakage through the left abutment grout curtain can be seen as wet spots on the face of the dam.

A new approach using a combination of fast setting chemical grout to seal the leakage from the original relief drains, cementitious grout to form a new upstream curtain, and the re-establishment of downstream relief drains was designed in 1996 by the Corps with assistance from the Bureau of Reclamation. The work was performed under contract in 1997 by Layne Keller LLC., a joint venture of Hayward Baker Inc. and Layne Christensen. ECO Geosystems was a technical consultant on the project. This paper is a case study of the design and actual construction of the new remedial grout curtain at Dworshak Dam.

Introduction

Dworshak Dam, Figure 1, was constructed by the Corps of Engineers (Corps) on the North Fork of the Clearwater River, approximately 56 kilometers (35 miles) east of Lewiston, Idaho. Construction was completed and the dam put into service in 1972. The Corps remains the owner/operator of the dam. The dam has a structural height of 219 meters (717 feet), and is the highest straight axis concrete gravity dam in the Western Hemisphere. The dam crest is 1,002 meters (3,287 feet) long at elevation 492 meters (1613 feet) above sea level. Dworshak Dam is the third highest dam in the United States. The level of the reservoir behind the dam is controlled for flood control, power generation, fish migration and recreation.

Original Grout Curtain and Relief Drains

The dam was constructed in a narrow valley with very steep side slopes. This paper will discuss the left abutment of the dam, monoliths 1 through 27, and specifically monoliths 15 through 19. Review of the Dworshak Dam Foundation Report (U.S. Army Corps of Engineers, 1971) indicates that the bedrock foundation in this area was composed of competent granite gneiss with structural orientations of 15 to 30 degrees dipping generally to the west. Features that were nearly vertical and striking northeast to southwest were also present. Inspection of the foundation from the dam adits revealed very competent rock that was slightly to very slightly fractured and jointed with widely scattered shearing. The fractures/joints were commonly infilled with clay and mica.

Foundation permeability as determined by pressure testing in bore holes, was moderate to very low, progressively decreasing with depth:

Depth		Permeability	
meters	(feet)	cm/s	(ft/min)
9.1	30	1.1×10^{-3}	2.2×10^{-3}
36.6	120	4.8×10^{-4}	9.5×10^{-4}
76.2	250	5.1×10^{-7}	1.0×10^{-6}

During dam construction, a single line grout curtain was created from the foundation and grouting gallery. The left abutment grout holes were drilled on 1.5, 3, and 9 meter (5,

Figure 2. High water flows in the grouting gallery caused personnel safety hazards and could have caused unsafe uplift pressures if curtain leakage exceeded relief drain capacity.

10 and 30 foot) centers perpendicular to the Construction Base Line (CBL) and at a slope of 3 vertical to 1 horizontal in the upstream direction, except in monoliths 14, and 15, where the grout holes were drilled vertically to cause a more effective cutoff for the diversion tunnel which ran under monoliths 13 and 14, and in monolith 16 where there was a transition section from vertical back to 3 vertical to 1 horizontal. Grout hole depths varied from 36.5 meters (120 feet), for holes on 1.5 meter (5 foot) centers, to 76.2 meters (250 feet), for holes on 9 meter (30 foot) centers. The grout injection pressures used to construct the original grout curtain were relatively low, commonly less than 6.9 kPa (1 psi) gauge per foot of depth.

A drainage curtain was constructed downstream of the grout curtain to reduce uplift pressures beneath the dam. These holes were NX size and drilled at a 3 vertical to 1 horizontal slope downstream and perpendicular to the CBL. Drainage curtain holes were located on 1.5 meter (5 foot) centers and commonly were drilled to a depth in excess of 53 meters (175 feet), in some instances exceeding the depth of the grout curtain holes.

Problem Flows From Relief Drains and Monolith Joint Drains

Seepage flows at full reservoir elevation from the left abutment drains were relatively constant until mid-1984 with flows from the drains in monoliths 14 to 17 totaling about 2,300 liters/min (600 gpm). After 1984 a significant increase in rate of flow began. Water production in 1987 had increased to 4,542 liters/min (1200 gpm), and measurements taken in mid-1996 revealed a total flow of 9,464 liters/min (2,500-3,000 gpm) from the drains in monoliths 14 to17. More than half that total was coming from the drains located within monoliths 16 and 17. The flow from the drains was visibly clear but observation of the various collection flumes showed that fracture infill material was being eroded. Although foundation uplift pressures remained well below the original design assumptions, there was concern that if flows increased beyond drain capacity an increase in uplift pressures could occur. Additionally, such flows exceeded the capacity of the left abutment drainage gallery, overtopping stairs, landings and gallery walkways causing personnel and safety concerns, Figure 2.

The monolith joints were constructed to form contraction joints. Two vertical 203 mm (8 inches) diameter formed joint drain holes were constructed in each monolith joint 2.75 meters (9 feet) and 5.5 meters (18 feet) from the upstream face. The monolith joints were provided with vertical bi-fold copper waterstops located 1.2 meters (4 feet) and 4 meters (13 feet) from the upstream face with the upstream waterstop embedded 0.3 meters (1 foot) into the rock foundation. Each pair of vertical monolith joint drains terminated in a horizontal drain that daylighted in the grouting and drainage gallery. The drains of monolith joints 14/15, 15/16, 16/17, and 17/18 showed significant leakage directly through the waterstops possibly due to waterstop failure or improper installation/concrete consolidation during construction.

Numerous investigations/evaluations of the problem water flows were performed during the period 1984 through early 1995. The overall conclusions from the investigations were that:

1. The flow was coming from fractures intercepted by the left abutment foundation drains in monoliths 15-19. These fractures were interconnected and drain hole cross communication was common.

2. Some individual drains had flows as high 757 liters/min (200 gpm) and pressures as high as 690 kPa (100 psi). Most drain hole flows, however, were significantly less than 378 liters/min (100 gpm) and pressures were below 276 kPa (40 psi).

3. Vertical features of the foundation may not have been intercepted and grouted during construction of the original grout curtain.

4. Some clay infill material from the foundation fractures was being piped into the grouting gallery.

First Attempt at Remedial Grouting

In the fall of 1995 a contract was let to perform remedial cementitious grouting in monoliths 14-19 from the grouting gallery. The reservoir elevation was lowered approximately 30 meters (100 feet) below full stage to elevation 457 (1500 feet) at the start of the grouting work.

During the very early stages of this program it became apparent that interconnections between the existing relief drains and the new grout holes would ultimately result in grout filling and closing the existing relief drains. The Corps thus decided to intentionally grout all the relief drains in monoliths 15 through 18 using a grout having a water/cement ratio (w/c) of 1:1 and lost circulation materials with cement accelerators. This was to be followed by grouting a new, remedial upstream curtain and drilling of replacement downstream drains. It was determined that foundation uplift pressures would remain within design considerations, even with the existing relief drains sealed, provided reservoir levels below elevation 457 meters (1500 feet) could be maintained. While the required materials were being gathered for this effort, the region experienced abnormally high rainfall. The reservoir level rose to elevation 466 (1530 feet) in spite of Corps' efforts to maintain elevation 457 (1500 feet). It was determined that there could be high spring runoff flows before the work could be completed, and with such high flows it might not be possible to maintain the reservoir elevation below that which might generate unsafe uplift pressures. The Corps thus elected to terminate the grouting work and reopen all relief drains.

Design of the Second Remedial Grouting Program

Using information and experience gained during the first remedial grouting program and with technical assistance of the U.S. Bureau of Reclamation, the Corps of Engineers designed a second grouting program for construction during the period May through December of 1997. This program consisted of three objectives:

1. Installation of additional dam instrumentation to continuously monitor dam uplift pressure and leakage flows and the setup of an instrumentation database to manage the instrumentation data generated during and after remedial grouting.

2. Remedial foundation drilling and grouting in monoliths 15 through 19 from the drainage and grouting gallery.

3. Repair leaking upstream monolith joint waterstops by grouting, including reestablishing any grout sealed downstream monolith drains.

It was decided to utilize the knowledge and experience of a speciality grouting and drilling contractor during this construction and a set of technical specification were prepared requesting proposals under Federal contract negotiation procedures.

The contracting procedure utilized on the project was a first for both the successful contractor and the Corps. The procedure required submission, for review by the Corps, of a detailed technical proposal and cost proposal that addressed Corps specifications provisions.

The specifications (U.S. Army Corps of Engineers, 1997) contained the following general provisions to accomplish the above remedial grouting program:

1. Furnish and install additional and replacement uplift pressure instrumentation, crack/joint displacement meters, seepage flow monitoring instrumentation and open tube piezometers complete with a database monitoring system capable of presenting the instrumentation data in spreadsheet form.

2. Remedial Foundation Grouting

a. Perform remedial grouting using 45,425 liters (12,000 gallons) of fast setting chemical grout, injected into existing foundation relief drains, to construct a temporary downstream curtain.

b. Drill, and grout with cementitious grout, a multi-row permanent upstream grout curtain including 6,096 linear meters (20,000 linear feet) of grout hole drilling.

c. Drill 2,469 linear meters (8,100 linear feet) of relief drain holes to establish a new row of downstream foundation pressure relief drains to replace those grouted during formation of the temporary grout curtain.

3. Controlling Monolith Joint Leakage

a. Install packers in leaking monolith joint drains to reduce and/or control leakage to less than 38 liters/min (10 gpm). (This work was specified because testing had verified that there was connection between the leaking relief drains and the foundation bedrock joints and fractures).

b. After completion of curtain grouting, repair upstream leaking monolith joint waterstops by grouting the upstream vertical drain holes.

c. After grouting the upstream vertical monolith drains, reestablish the downstream monolith joint drains by cleaning, or by drilling replacement downstream joint drains.

The work to be performed in monoliths 15 through 19 involved extremely difficult working conditions in the steeply sloping 1.8 meter (6 feet) wide by 2.4 meter (8 feet) high drainage and grouting gallery. The gallery was congested with stair-steps, handrails, and a multitude of drain pipes and other plumbing. Total water leakage from this gallery was in excess of 15,100 liters/min (4,000 gpm) at an average temperature of 4.4° C (40° F). The time frame, May 15 through July 15, 1997, was established as a mobilization and instrumentation installation period while the Corps was lowering the reservoir water level to elevation 466 meters (1530 feet). The time frame July 15 through December 15, 1997, was designated as the monolith joint sealing and remedial grouting period while the reservoir level was maintained at elevation 466 meters (1530 feet), or lower if possible. A construction completion date of December 15, 1997 was required in order to allow reservoir filling for fish mitigation and recreational purposes.

Upon award of the contract, the technical proposal submitted by the contractor was integrated with the Corps specifications and pricing into a final contract document. Differences in the specified method of work and contractor proposed methods of completing the work were identified during a preconstruction meeting and an exact definition of work activities was determined and included as an addendum to the contract. The contract was performed on a formal "Partnering" basis allowing any required changes to the contract procedures and or performance to be identified quickly and made as needed.

Field Applications of Remedial Grouting

Dworshak Dam is located approximately 3.2 kilometers (2 miles) upstream of a national wildlife fish hatchery. Extreme care had to be taken by the contractor to insure that no potential pollutants including petroleum products, cement, polyurethane grout waste, and other potential hazardous wastes were allowed to enter the surface and subsurface waters. An extensive environmental protection program was instituted prior to construction and maintained during the project.

A network of berms, dikes, and ponds was utilized to settle out turbidity causing materials. An existing unused powerhouse area (skeleton bay) was used to minimize turbidity and was designed to be used as a chemical treatment area to control excessive turbidity resulting from the drilling and grouting operations. A high volume submersible pump was used to pump the treated waste water from the skeleton bay through an unused diversion tunnel where further environmental treatment could be performed if necessary.

The environmental protection program resulted in no turbidity increases above background levels being encountered during the entire project.

An extensive Contractor Quality Control (CQC) Plan was implemented on the project. All construction quality control, including the implementation of any construction activities that differed from the specified procedures, was the responsibility of the CQC manager. CQC requirements included integration of the environmental plan, operational plan, instrumentation monitoring, and distribution of information. Instrumentation installation and monitoring in accordance with the CQC Plan was also an important and necessary requirement of the contract.

Instrumentation Installation

The contract required grouting of all existing relief drains and sealing of the monolith joint drains prior to remedial curtain grouting. This had the potential for creating excessive foundation uplift pressures. Installation of an extensive instrumentation system was specified to monitor uplift pressure, foundation head pressure, and gallery water flow. Pressure transducers were installed into two existing and five new open tube piezometers, which were constructed using diamond core drilling techniques to a depth of 76 meters (250'). Pressure transducers were also installed at crack meter locations to monitor potential flows through cracks in the concrete that had developed during and after construction of the dam. A bulkhead and flume, with remote access telemetry information, was installed along the gallery floor to monitor gallery flows up to 11,400 liters/min (3,000 gpm). All instrumentation information was monitored with a laptop computer located in the gallery area to provide immediate access to the information. Instrumentation information was continuously monitored and reported to the Corps on a daily basis. The instrumentation system was left in place at completion of the project for continued use by the Corps.

Monolith Joint Drain Sealing and Repair

The contract required sealing of the monolith joint drains that exited into the gallery and controlling the leakage from the monolith joints to reduce flows to a maximum of 38 liters/min (10 gpm) per drain. Extensive leakage, as much as 50 percent of the total flow in the gallery, was experienced through the monolith joints and joint drains. The contract also required maintaining the vertical downstream drain, and the horizontal drain section in each monolith as a functional drain while grouting the upstream drain for use as a waterstop because of the failure of the original copper waterstops in the dam. Several of the monolith joints leaked from old drill holes that penetrated through the gallery waterstop. Other joints leaked directly through the waterstop due to waterstop failure or improper waterstop installation during dam construction. Inspection of the joint drains using an underwater camera was performed prior to commencement of sealing and at the completion of the grouting operations to determine the condition of the monolith joint drains and to determine if obstructions existed. Sealing of upstream monolith joints was specified to be performed with a flexible hydrophobic MDI based, solvent free, water reactive polyurethane grout. The technique of temporarily sealing the downstream joint was to be with a method proposed by the contractor.

Based on information obtained during the grouting operations, it appeared that the monolith joints drains were connected to the foundation fracture system. The contractor determined, and proposed, that treatment of the fractures below the monolith joint and the waterstop prior to sealing the gallery relief drains would result in a substantial reduction of water flow into the gallery through the relief drains.

Leakage into the gallery from the monolith joints and existing exploration holes was treated using polyurethane grout injection behind the water stop and directly into the existing holes to seal the holes and the drains. In this work, a 95mm (3/8") diameter hole was drilled on an angle from the gallery behind the joints at several points, and directly into the exploratory holes. Polyurethane resin was injected through grease nipple injection ports (zerks) inserted into the drill holes and the leakage was eliminated.

The specifications required sealing of four upstream monolith joints, 14/15, 15/16, 16/17, and 17/18, with flexible polyurethane grout to act as a permanent waterstop. The contractor's technical proposal included a procedure for polyurethane grouting of the foundation rock and rock/concrete contact area through the monolith joint drain prior to sealing the existing relief drains to reduce the flows to gallery relief drains. This proposal was accepted and was performed as follows:

- Video inspections of the 203 mm (8 inches) diameter drain holes were conducted using an underwater camera.
- Temporary 152 mm (6 inches) diameter casing was installed to the external radius of the elbows in the cleaned out joint drains to act as a guide for subsequent drilling operations.

- One hundred fifty two, 152 mm (6 inches) diameter holes (rat holes) were drilled vertically through the elbows of the joint drains, past the rock/concrete contact, and 3 meters (10 feet) into bedrock to allow access to the foundation fractures connecting to the joint drain. Drilling was conducted with a truck mounted top drive rotary drill, using downhole hammer percussion drilling with air circulation in the concrete and rock portions of the hole The vertical depths of the monolith joint drains varied from 106 to 122 meters (350 to 400 feet) from the dam surface to the bottom of the drilled hole and required the use of a pressure booster air compressor to evacuate drill cuttings and water from the hole..
- A 50 mm (2 inch) diameter, Schedule 80, PVC multi-port sleeve pipe (MPSP) packer/grouting system was installed into each drilled hole from the surface of the dam to the bottom of the hole. The MPSP system, commonly known as a tube-a-manchette or sleeve port grout pipe system, utilized one way grout ports installed on 610 mm (2 inch) centers along the MPSP pipe. The MPSP pipe had geotextile bag packers (barriers) attached to the pipe, straddling sleeve ports, at strategic locations along the pipe. The barriers served to isolate zones along the borehole to allow chemical grouting under pressure in the isolated zones. The barrier bags were inflated using bentonite/cement grout mixture to seal off the zones.
- Chemical grout was injected through a specially designed pneumatic packer inserted into the PSP pipes at various locations using 19 mm (3/4 inch) diameter supply hose for polyurethane resin and 6.4 mm (1/4 inch) tubing for reactive water. Water-resin mixing occurred within the pneumatic packer. Grouting was conducted by zones from the foundation rock contact to the roadway surface of the monolith joint drains.
- Chemical grout was placed using a rocker pump with a capacity of 25 liters/min (6.5 gpm) and 28 Mpa (4,000 psi) pressure at the pump. The pump was capable of supplying water to resin ratios varying from 1:1 to 20:1.

The downstream monolith joint drains were required to be maintained as functional joint drains after completion of the drilling and grouting operation. The extent of water inflow (and possible cement grout flow) into the drains was unknown, but was anticipated, and thus required the use of a temporary sealing procedure. A 50 mm (2 inch) steel sleeve pipe with injection ports on 1.5 meter (5 foot) centers along the pipe, and a woven geotextile bag, double layered which enveloped the sleeve pipe was used because the steel pipe and bag could be removed after grouting. The geotextile bag and pipe was installed into the hole to the elbow of the joint drain and bentonite slurry was injected into the bags in sections starting at the bottom and working up using a straddle packer and injection pipe inside the sleeve pipe. The horizontal section of the monolith joint drains exiting in the gallery were sealed using a gasketed plate bolted to the gallery wall face with a pressure relief valve attached to monitor flows.

Very high volumes of water at high pressures were encountered during drilling of the rat holes. Grouting of monolith 15/16 was completed using polyurethane grout and monolith 14/15 was partially completed using polyurethane grout. The high water velocity through the monolith joints caused the injected resin to wash out through the downstream face of the dam prior to the grout curing. After substantial effort at treating the monolith joints with polyurethane methods failed, an alternative method of monolith joint sealing was developed cooperatively under the partnering provisions, accepted by the Corps, and performed by the contractor.

Alternate Monolith Joint Drain Treatment

The alternative method of sealing the monolith joints was to first use bentonite chips and sawdust mixed in a bentonite gel slurry and pumped into the downstream monolith joint drains with a concrete pump to form a temporary downstream seal. After the downstream seal was accomplished, a permanent cementitious grout mixture, designed to have a final strength less than that of the mass concrete to facilitate re-drilling if required, was used to grout the upstream monolith joints. After placement of the upstream joint grout, the temporary downstream bentonite seal was washed/drilled from the vertical section of the downstream joint drains and ratholes. Permanent cementitious grout was then pumped into the portion of the ratholes located in rock to seal foundation rock fractures. The function of the downstream joint drains was thereby reestablished. Accomplishment of this alternative involved:

- Removal of the MPSP pipe from the monolith joint drains. The steel pipe was pulled from the downstream joints and the PVC sleeve pipe and chemical grout was drilled out of the upstream holes using rotary drilling methods with air circulation.
- Drilling of 152 mm (6 inch) diameter rat holes 3 meters (10 feet) into rock in the downstream joint drains similar to the upstream drains.
- Onsite mixing of the bentonite chip, sawdust, and bentonite gel mixture and pumping into the downstream joint drains and ratholes using a concrete pump.
- Cementitious grout mixture batched off site at a ready mix facility and transported to the site for placement with a concrete pump. Grout was placed with a 102 mm (4 inch) diameter tremie pipe from the bottom of the hole upward.
- Cementitious grout mixture proportions (per cubic meter)

Cement	178 kg (392 lbs)	Air entraining agent	889 ml (30 oz.)
Flyash	356 kg (784 lbs)	Water reducing agent	4,259 ml (144 oz.)
Water	237 kg (523 lbs)	Anti washout agent	5,416 ml (183 oz.)
Sand	1,246 kg (2,747 lbs)		

The Corps specified that the cementitious grout have 28 day compressive strengths between 1.38 and 13.8 MPa(200 psi - 2,000 psi).

Because of the critical schedule for installation of the remedial grout curtain, sealing of the monolith joints was performed concurrently with the curtain grouting operation. This resulted in some of the horizontal sections of the monolith joint drains becoming filled with curtain grout. Horizontal 254 mm (10 inch) diameter holes were drilled from the gallery, along the affected monolith joints, to intersect the downstream joint drain to allow water to drain into the grouting and drainage gallery. These drill hole lengths varied from 1.2 to 5.2 meters (4 to 17 feet) and were completed using thin wall diamond core drilling methods with electric powered stand drills.

Because of the successful sealing of the left abutment monolith joint waterstops using the above described alternate method, the Corps issued a change order to perform similar treatment of an additional right abutment monolith joint that was experiencing high water flows. All monolith joint sealing and repair was completed within the original contract schedule period.

Foundation Drain Holes

Sixty three existing NX size foundation drain holes, with depths to 69 meters (225 feet), required sealing by polyurethane chemical grouting prior to curtain grouting operations to minimize water flows through the foundation and allow successful grout curtain installation. Some of the drain holes had water flows in excess of 84 liters/min (100 gpm), with total flows through the foundation drains estimated at approximately 1,600 to 2,000 litres/min (2,000 - 2,500 gpm). The contractor proposed a chemical grout solution utilizing a 51 mm (2 inch) diameter MPSP packer system to facilitate grouting of each hole while cutting off the water to the gallery with barrier bags and helping to insure that only localized zones in adjacent holes, and not the entire hole, was grouted by communication.

Grout delivery system
• The MPSP pipe consisted of 51 mm (2 inch) inch inner diameter, PVC pipe, with a bottom cap to limit water intrusion into the pipe, with a number of exterior barrier bags strapped to the sleeve pipe to provide grout zones. Once the MPSP pipe was installed in the foundation drain holes, the barrier bags were inflated with bentonite/cement grout.
• Zones were water tested using a double packer inside the MPSP pipe to determine flow rates.
• Injection of polyurethane grout was through a single pneumatic packer arrangement inserted into the MPSP pipe. The packer was placed at the lowest zone, below the lowest barrier bag, and grouting was performed using upstage methods.
• The packer line contained 2 concentric tubes; a 19 mm (3/4 inch) tube for placement of polyurethane resin, and a 6.35 mm (1/4 inch) nylon tube for placement of reactive water. Water-resin mixing occurred within the pneumatic packer.
• The entire sleeve pipe below the grout packer, with the numerous one way valves, acted as a mixing chamber for the resin and reactive water.
• Polyurethane grout was placed using a rocker pump with 25 liters/min (6.5 gpm) and 28 Mpa (4,000 psi) capacity.

• Hydrophobic polyurethane resin grout was used.
• A significant laboratory testing program was conducted to determine reaction rates of the polyurethane grout under high water velocities and cold water conditions prior to implementation on the project.

Placement of the MPSP and inflation of the barrier bags with bentonite/cement grout sealed off interconnected water flows on many of the foundation drains. High water velocities in the foundation fractures and very low water temperatures limited reaction of the polyurethane resin in many of the foundation drains. Grout takes in other holes was lower than had been anticipated because the MPSP cut off high velocity water flows into the drains. The installation of the MPSP in low water flow drains resulted in the water flows being re-channeled into other drains and causing an increase in flow from drains in which MPSP pipe had not yet been installed. In some instances, substantial flows prevented the installation of MPSP in drain holes and necessitated a change in grouting methods. Based on discussions with the Corps, cement grout with sodium silicate accelerator was utilized to grout high flow drain holes where polyurethane grout was not effective, or where the MPSP could not be installed. Extensive communication from the foundation drains to intersecting gallery cross drains, other gallery drainage systems, and the downstream face of the dam occurred. Various methods, including grouting with and without mechanical packers, were used to seal the associated drainage systems and allow successful sealing of the foundation drains.

Approximately 10% of the foundation drains were grouted with polyurethane resin using 1,484 liters (392 gals.) of resin (approximately 3 % of the contract estimate). An additional 898 bags of cement grout were utilized to seal the high flowing drains. All of the drains were successfully sealed within the 30 day contract schedule at a contract cost of approximately 30 percent of the anticipated amount. Water flows from the foundation relief drain holes were thus reduced to essentially zero.

Remedial Grout Curtain Installation

A two row remedial grout curtain, upstream of the original grout curtain, was required to be installed after the existing gallery relief drains were grouted and sealed. All remedial grout curtain holes were 40 meters (130 feet) deep. The upstream row was located in the upstream gallery ditch with the downstream curtain row located two feet downstream of the upstream curtain row and within the alignment of the gallery stairs. Curtain grout holes were spaced 1.5 meters (5 feet) apart. Grout holes were oriented 20 degrees and 60 degrees from vertical and angled into the abutment. Approximately 85 holes were required for the upstream curtain row and 77 holes were required in the downstream curtain row. Actually, the existing grouted relief drain holes formed a third and most downstream row of the remedial curtain.

Grout hole drilling was accomplished using screw feed diamond rotary core drills mounted on column posts held in place on the invert and roof with jacks. Formations to be drilled included high strength concrete and hard granite rock formations. High speed, electric-hydraulic diamond drills were anticipated to be the primary drill for the hard formations but could not be utilized because of the confined area, inability to move equipment quickly, and safety concerns. Grout holes were required to be a minimum of 48 mm (1.89 inches) AX size and were required to be cored. Height limitations required that 610 mm (2 feet) maximum rod lengths could be used. Thin kerf AW-34 conventional coring systems with extendable core barrels and impregnated diamond core bits 48 mm x 33.5 mm (1.89-inch OD x 1.32-inch ID) were used to maximize production.

An average of five drills, working 24 hours per day, 5 days per week, were used to complete drilling of the grout and relief drain holes within the required contract schedule. After an initial sequencing delay resulting from higher than anticipated water flows and relief drain grouting delays, 6,050 meters (19,848 feet) of grout holes were drilled with an average production rate of approximately 17.5 meters (57.4 feet) per shift per drill.

The original specified procedure was upstage grouting, with occasional downstage grouting whenever drill water loss occurred, or whenever bad hole conditions were encountered. As part of the technical proposal the contractor identified that most of the water would probably be encountered in the top portions of the holes and proposed a combination downstage/upstage process which was accepted by the Corps. The revised procedure significantly reduced the number of hookups and pressure tests that were required to be completed.

- The upstream grout holes were grouted in two stages. The top 18.3 meters (60 feet) were grouted with a stage down method. If substantial water loss did not occur, the hole was grouted in one 18.3 meter (60 foot) stage from the collar of the hole. If water loss occurred, the 18.3 meter (60 foot) stage was grouted on a stage down basis as needed. All of the top stages in an area were grouted to completion prior to extending the holes to total depth to maximize water cutoff.
- After completion of the top 18.3 meter (60 foot) stage the holes were drilled to total depth and upstage grouted from 39.6 meters (130 feet) to 18.3 meters (60 feet).
- The downstream curtain holes were grouted using upstage methods in accordance with the specified procedure except that drilling was stopped at the 18.3 meters (60 feet) depth and water pressure tested to determine flow rates for the top 18.3 meters (60 feet). The holes were then continued to total depth.
- Grout materials were mixed at an outside batching location and delivered approximately 370 meters (1,200 feet) by pipeline to the gallery area. Grout materials were delivered by using compressed air to push the mixed grout to transfer/placement tubs located within the gallery.
- The grout mixer/agitator was an air powered one third cubic meter high speed colloidal type mixer with diffuser-type centrifugal mixing pump and a one third cubic meter agitated holding tank.

- Transfer tubs were also one third cubic meter agitated holding tanks equipped with Moyno 3L-10 helical screw pumps for grout placement. Three transfer tubs were located within the gallery near the grout curtain area to minimize grout pumping distances. Due to the confined locations specialized transfer tubs were manufactured specifically for the project.
- Grouting operations were conducted using 1 to 2 pumps per shift depending on grout hole availability.
- Grouting of the remedial curtain holes was accomplished with cement grout using admixtures for acceleration and high range water reducing admixture (HRWRA). Thirty two hundred bags of cement (65 percent of anticipated quantity) were placed with an average production rate of approximately 55 bags per shift. HRWRA quantities used were 36 percent of estimated and accelerator was 113 percent of estimated quantities. A substantial quantity of accelerator were used in the cement grouting of the existing relief drains and to accelerate grout used in sealing other gallery drainage structures. Grout w/c ratios varied from 5:1 to 0.7:1. Water testing of grout holes was used to determine starting grout mixes. Starting grout w/c ratios were generally 3:1 with reduction to 1:1 and occasionally to 0.75:1 at closure. Overall average grout mix ratios were approximately 1:1.

Reestablishment of Relief Drains

The existing foundation drains had hole depths up to 69 meters (225 feet). It was determined by the Corps that foundation water flows were going below the shallower grout curtain and entering the deeper foundation joint drains. A replacement foundation drainage system was installed after completion of the remedial grout curtain. The new drainage system utilized 76mm (3 inch) diameter holes located on 1.5 meter (5 foot) centers between the grouted existing drain holes and drilled to a depth of 30 meters (100 feet). Holes were drilled using diamond core drilling methods utilizing NX size drilling bits, drill rods, and core barrels. Narrow kerf 1.5 meter (5 foot) wireline core barrels and impregnated diamond bits 76 mm OD x 51 mm ID (3 inches x 2 inches) were used to maximize drilling production. Drill rigs were the same as used for grout hole drilling. Actual drain footage was 1,697 meters (5,567 feet) (55% of anticipated) with average production per drill shift of approximately 11 meters (36 feet).

Summary

The instrumentation installation, monolith joint sealing, existing relief drain grouting, grout curtain installation, and new relief drain installation at Dworshak Dam was required to be performed in a very short time schedule and under difficult access and water conditions. During initial performance of the gallery setup, grout nipple installation, and installation of the MPSP system in the gallery relief drains, the contractor was working with more than 15,000 liters/min (4,000 gpm) of water flowing through the gallery work area. After installation of the MPSP and sealing of the monolith joints the water flows were reduced to 750 - 1,100 liters/min (200-300 gpm) and final flows upon completion of the grouting operations in December of 1997 were in the order of 95 litres/min(25 gpm), significantly less than the 7,500 - 11,500 litres/min (2,000-3,000 gpm) normally associated with similar reservoir level elevations. On June 1, 1998 the reservoir elevation was .3 meters (1 foot) below full pool elevation and water flows through the gallery were less than 1,100 liters/min (300 gpm). According to Corps of Engineers sources this is the lowest flow rate through the gallery since the dam was constructed.

The original contract required completion of gallery work, 8,535 linear meters (28,000 feet) of hard rock grout and drain hole drilling, placement of 45,425 liters (12,000 gallons) of chemical grout, injection of 4,000 bags of cement grout within a very tight schedule of approximately 150 calendar days. The work was completed ahead of schedule and the balance of the project was completed more than a month ahead of schedule including an additional 7 percent of the original contract value added as change orders.

The formal partnering procedure was successfully used by the Corps and the contractor and allowed for technical completion of the project at a cost of approximately 80 percent of the original contract value including change orders for extra work added to the contract. Turn around time for change orders necessitated during the work was less than 2 weeks.

REFERENCES

U.S. Army Corps of Engineers, "Dworshak Dam and Reservoir, "Foundation Report, Chapter 3 - Dam and Powerhouse", Walla Walla District Office, Walla Walla, Washington, 1971.

U.S. Army Corps of Engineers, "Request for Proposal, DACW68-97-R-0004, Curtain Grouting Monoliths 15 through 19 Dworshak Dam, Clearwater County, Idaho", Walla Walla District, Walla Walla, Washington, January 1997.

Grouting of TBM Rock Tunnels for the Los Angeles Subway

Gary Kramer[1], Michael Roach[2], John Townsend[3] and Stuart Warren[4]

Abstract

Twin subway tunnels, each 4 km (2.5 miles) in length were TBM mined in urban Los Angeles through extremely varied geologic conditions. Ground consisted of alluvial soil, shale, sandstone, basalt and granodiorite rock and varied from massive to blocky, shattered to completely decomposed to fault gouge material. TBM advance was possible by the implementation of various grouting methods for ground improvement, water inflow control, environmental issues and settlement limitation.

Introduction

Contract C0311 of the Los Angeles County Metropolitan Transportation Authority's (MTA) Red Line subway expansion called for mining twin tunnels (AR and AL) in rock each 4042m (13,260') in length and 6.3m (20'8") in diameter. This work is currently being completed by a joint venture (TBI/FKCI) of Traylor Bros., Inc. and Frontier Kemper Constructors, Inc. both from Evansville, Indiana. These tunnels beneath the Santa Monica Mountains will provide a heavy rail subway link between the Los Angeles Basin and the San Fernando Valley for the citizens of Los Angeles.

The tunnels extend south from Universal City Station in the Valley to junction with previously mined soft ground tunnels near Hollywood Boulevard (see Figure 1). The tunnel section designer is Parsons Brinckerhoff and construction management is provided by JMA, a joint venture of Jacobs Engineering Group of Pasadena, CA, Hatch Mott MacDonald of Pleasanton, CA and ACG Environments of Los Angeles.

Geologic Conditions

The tunnel passes through a truncated anticline of Cretaceous and Palaeocene rocks in the Santa Monica Mountains bounded by the Benedict Canyon Fault to the north and truncated to the south by the active Hollywood Fault.

---

[1] Supervisory Engineer, Hatch Mott MacDonald, Pleasanton, California
[2] Project Manager, Traylor Brothers Inc., Evansville, Indiana
[3] Associate, Hatch Mott MacDonald, Pleasanton, California
[4] Specialist Geologist, Hatch Mott MacDonald, Pleasanton, California

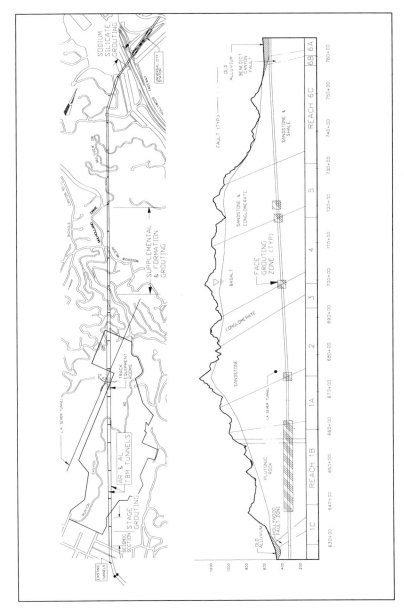

Figure 1: Plan View of Alignment and Geologic Profile of C0311 Tunnels

'Old' alluvial deposits are present north and south of the mountains. Figure 1 and Table 1 briefly summarize the Geotechnical Design Summary Report's (LACMTA, 1994) description of the ten geologic reaches lying along the tunnel alignment:

Formation	Reach #	Description	Permeability (cm/sec)
Upper Topanga	6A (360')	Starter Tunnel: weathered weakly cemented, clay rich sandstone & shale	5.2×10^{-8} to 4.2×10^{-4}
	6B (170')	Benedict Canyon Fault: Gouge and distorted shales and sandstones	
	6C (2460')	Weakly cemented, clay rich sandstone and shale	
	5 (1460')	Moderately well cemented sandstone and conglomerate	
Middle Topanga	4 (1780')	Highly fractured basalts and basalt breccias with sandstone beds	8.0×10^{-7} to 3.2×10^{-5}
Lower Topanga & Las Virgines Sandstone	3 (700')	Massive sandstones	$<3.3 \times 10^{-7}$ to 4.7×10^{-7}
Simi Conglomerate & Chico Sandstone	2 (1100')	Massive sandstones and conglomerates	2.9×10^{-6}
Upper Cretaceous Plutonic	1A (2400')	Moderately fractured Quartz-Monzonite and Granodiorite	4.6×10^{-8} to 1.7×10^{-4}
	1B (1960')	Highly fractured granodiorite with numerous minor fault zones	
	1C (870')	Special Seismic Section with fault gouge and intact blocks of granodiorite	

Table 1: Description of Geologic Reaches along Tunnel Alignment

Hydrostatic heads up to 27 bar (400psi) occur along the tunnel alignment at discrete hydrogeological units bounded by numerous faults acting as aquitards. The Hollywood Fault forms a major hydrogeologic boundary with a 60m (200') differential head across the fault and between the alluvium and granitic rock. Throughout the length of the alignment, numerous shears, faults, fractured and blocky rock and gouge zones were anticipated that could affect tunneling progress.

Construction records from the nearby 1955 Los Angeles Sewer Tunnel demonstrated that the basalt and granitic rocks had a high secondary permeability with fissures and contact zones along faults and between different rock types forming the primary flow paths.

Tunneling Methods

The contract specified that two TBM's simultaneously mine the twin tunnels. Consequently, TBI/FKCI selected two similar Robbins main beam machines which combined had mined over 22,900m (75,000') of tunnel on previous projects. Construction & Tunneling Services, Inc. (CTS) of Kent Washington remanufactured both TBM's for this contract. The photograph in Figure 2 shows the AR tunnel machine assembled in the Universal City Station excavation prior to excavation.

Figure 2: AR TBM "Thelma" Ready for Launch

Although quite similar in appearance, the machines were not identical twins. Modification work was undertaken to change the bearing, bull gear and conveyor configuration to upgrade and match the two machines. This was considered essential to provide adequate bearing life at full thrust and allow a common set of bearings and gear spares. Changes to the TBM's included:

- Installation of high torque, variable speed hydraulic drives with full reverse capacity replaced direct electric drives.
- New back loading, flat profile cutterheads to support 432mm (17") single and double CTS disc cutters with 225kN (50,000 lb) minimum thrust capacity each.
- The roof shields were reconfigured to maintain a close support surface to the cutterhead and provide 0.85 tunnel diameters of overburden support.
- Enlargement of the gripper shoes and addition of a construction notch to step over rib steel. For soft formations, the TBM hydraulics allow gripper pressure to vary with forward thrust pressures.
- Installation of an auxiliary invert ram system providing a secondary thrust off the invert segment to be used in the event of gripper slippage and assist in up steering in soft formations.

Prior to TBM delivery, MTA directed that the TBM probe and grout hole drilling capacity be upgraded from one to three face drifters due to growing concern regarding groundwater inflow levels and the possibility of additional grouting.

The contract allowed for several types of initial support to be followed by a cast-in-place permanent concrete lining but a mainbeam TBM does not permit efficient dowel installation or the use of full circle pre-cast segments. TBI/FKCI proposed all initial support to consist of a steel rib resting upon a precast invert segment with rock between ribs exposed but supported by steel mat lagging.

In keeping with deep-rooted tradition and a Hollywood theme, the two machines were renamed following a competition when one TBM became Thelma and the other became Louise.

Ground Stabilization by Chemical Grouting

The tunnels commence from the Universal City Station excavation and pass beneath the 10 lane wide Hollywood Freeway. This initial portion or starter tunnels lie within a mixed face of alluvial soil overlying soft rock. This condition, combined with the hard rock anticipated for the rest of the tunnels dictated that conventional methods be used for the short starter tunnels. In addition, an arched zone above the starter tunnels was permeation grouted with sodium silicate prior to excavation to minimize ground loss and reduce groundwater inflows (see Figure 3).

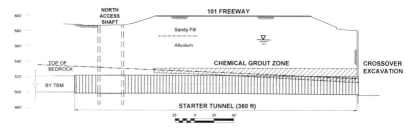

Figure 3: Profile through Starter Tunnels showing Chemical Grout Zone

Four levels of horizontal sleeve-port grout pipes (SPGP) spaced at 1.5m (5') centers were installed from a rectangular shaft excavation to provide grout zone coverage. Target injection volumes were established using estimates of ground porosity and grout travel. Grout was injected in three stages under a total volume limit and combined minimum pressure/flow criteria. Upon completion, over 3287m (10,780') of grout holes were drilled and over 2,650,260 liters (700,200 gal) of sodium silicate grout was pumped. Taylor et al (1997) and Kramer and Albino (1997) provide more detailed accounts of this grouting.

Upon completion of grouting, starter tunnel excavation commenced with the grouted ground remaining stable and with minimal groundwater inflows. The grouting effectiveness was evident as ungrouted starter tunnel portions experienced joint seepage and washout with loss of fines. This resulted in some face instability.

Not unexpectedly, some freeway settlement occurred. Average maximum trough settlement was 85mm (3.5") or 2.4% (percentage volume loss relative to the excavated tunnel volume) and includes combined effects due to dewatering, adjacent shaft excavations tunneling and consolidation. Average settlement from tunneling was only 50mm (2"). This settlement level is successful for conventional tunneling in general, but is particularly good for ground disturbed by previous shaft work and containing a large layer of fill material. The use of sodium silicate grout combined with good tunneling practice was responsible for this success. After starter tunnel completion to a full rock face, TBM's were moved up and mining commenced.

Consent Decree for Environmental Issues.

Environmental concerns arose regarding the potential for tunneling to impact surface springs and for groundwater depletion in the mountains. An MTA Board motion addressed these issues which were also incorporated into a Consent Decree in response to court action by environmental groups 18 months after the C0311 contract was let. The decree limited groundwater discharge to 13,400 m^3/day (3.75 mgd) and required the MTA "*to carry out whatever grouting is necessary, ahead of and behind the boring machines to ensure there is "zero leakage" beneath the seasonal springs*".

As these requirements were not included in the original design, a three phase grouting system was developed under change order to fulfill the requirements of the consent decree. This consisted of:

- Face grouting through probe holes drilled ahead of the TBM
- Supplemental Grouting behind the TBM's but prior to final lining installation
- Formation Grouting through the final cast-in-place concrete lining.

Springs were principally associated with faults or rock type boundaries. Consequently, seven 65m (200') wide grout zones were initially identified at geological features within the Upper Topanga sandstones, basalts and granitic rocks. The grout zone number and lengths were modified during tunneling in response to groundwater inflows and observation well monitoring.

Face Grouting

Original contract requirements included probing ahead of the TBM face to determine groundwater inflow potential and direct grouting when probehole inflows exceeded 175 l/m (50gpm) per 30m (100') of probe hole. In the Consent Decree grout zones, grouting was carried out when inflow exceeded 7 l/m (2gpm) per 30m (100'). Each TBM was equipped with 2 fixed hydraulic Boart rotary drifters above springline and a third drill was temporarily mounted below springline. These were subsequently supplemented with rotary-percussive drills in the hard granitic rocks.

Cutterhead limitations reduced the available grout pattern to 4 horizontal holes drilled through cutterhead buckets. To achieve desired grout travel, a fine ground slag cement (Fosroc 10-92 Ultracem with integral micro-dispersant) refered to as microfine was used except during major takes where Type III Portland cement was used. Often, Type III would meet with pressure refusal and a switch to microfine cement would result in significant grout take in the same hole. Volumetric water/cement ratios of 1.5:1 to 3:1 were mixed and pumped at the face to pressure refusal of 3.5 bar (50 psi) above hydrostatic head. Face grouting was carried out at 8 grout zones (see Figure 1) and total quantities are given in the following table:

Grout Hole Length Drilled	Microfine Cement	Type III Cement
1573m (5,159')	42,133 kg (92,940 lb)	42,491 kg (94,470 lb)

Table 2: Summary of Face Grouting Activity in AR and AL Tunnels

Verification holes were repeatedly drilled and grouted until the 7 l/min per 30m of hole (2gpm/100') criteria was met. Face inflow was generally less than 35 l/m (10gpm) per 1.2m (4') shove during excavation through the grouted zone.

Supplemental Grouting Prior to Tunnel Lining:

On several occasions, unacceptable groundwater inflow levels occurred after passage of the TBM primarily due to relaxation of the ground and incomplete annular face grouting due to cutterhead limitations on the grout pattern. In order to meet the groundwater discharge limits of the Consent Decree, a Supplemental Grouting program prior to final lining was developed and implemented (see Figure 1). This involved drilling rings at 2.4m (8') spacing of six radial holes each 3.6m (12') long and 50mm (2") in diameter (see Figure 4). Each hole was subsequently grouted with either microfine or Type III cement through mechanical packers at the hole collar. Refusal criterion was zero take at 14 bar (200 psi). Additional holes were added to the pattern holes to target specific fissures and inflows as required.

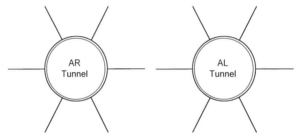

Figure 4: Supplemental Grout Hole Pattern

Grouting began in the basalts (Reach 4) and was surprisingly successful with no liner present to prevent communication. When communication did occur, fissure chinking was required. With relatively open fissures in the basalt, grout takes per hole would often exceed 100 sacks (13.6kg, 30lb each) of microfine, prompting implementation of a volume criterion of 30 sacks of microfine to determine a switch to Type III cement. Occasionally, the switch to Type III would result in pressure refusal within a few sacks. Table 3 provides total quantities to grout 600m (2000') of twin tunnel. Groundwater inflows were monitored in the invert to assess the effectiveness of the grouting. During grouting, the combined AR and AL inflows reduced from 1500 l/min (420 gpm) to 535 l/min (150 gpm) (see Figure 5).

Length of Grout Holes Drilled	Microfine Cement	Type III Cement
4883m (16018')	68,750 kg (151,590 lb)	134,625 kg (296,852 lb)

Table 3: Summary of Supplemental Grouting Activity in AR and AL Tunnels

Clay joint infill in the granodiorites (Reach 1A) inhibited grout travel and supplemental grouting was not continued as compliance with the total discharge levels was assured. It was felt that further flow reductions in this area would be best achieved with the 300mm (12") thick concrete final liner in place.

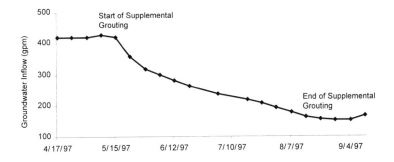

**Figure 5: Combined AR and AL Invert Flow in the Basalt (Reach 4)
during Supplemental Grouting**

Formation Grouting through the Tunnel Liner

Despite face and supplemental grouting efforts, several springs were observed to decline or dry up and observation wells showed water table drawdown. A water balance study of the area was carried out to identify water table recovery rates versus inflow levels into the lined tunnel. The study determined that long term inflow from the basalt and granitic rocks must be reduced to below 178 and 285 l/min (50 and 80gpm) respectively or 40% of the levels at completion of tunneling.

In the granitic rocks, liner sections were redesigned with higher strength concrete and without weepholes. This was not feasible in the basalts due to higher hydrostatic heads and a formation grouting program was developed to create a low permeability annulus around the tunnel. This work has yet to be carried out at time of writing. Radial ring patterns, on 2.4m (8') centers consisting of eight 3.6m (12') long holes are planned. As per supplemental grouting, microfine cement will be pumped to similar refusal criteria to further reduce annulus permeability. Formation grouting with the final liner in place is anticipated to be even more effective in reducing the permeability than that achieved in the supplemental grouting.

Tunnel lining strains will be monitored by strain gauge arrays to confirm the adequacy of allowable grout pressures. Following grouting, verification holes will be iteratively drilled and grouted until inflow criteria have been met. Piezometers installed up to 18m (60') outside the liner will monitor water table recovery.

Polyurethane Grouting

Polyurethane (PU) has been employed by the mining industry for nearly 30 years and surprisingly, its value in underground civil applications is not widely known in the United States. During excavation of Reach 5, loose ground was encountered in an unmapped shear zone in front of the AR TBM and a void developed halting excavation. From previous experience, TBI/FKCI suggested PU grout be used to fill this void and add cohesion to the loosened rock mass. Representatives of Micon, Grand Junction, Colorado, came to site and quickly developed a grouting method matching the C0311 geology and TBM equipment configuration. Over 2 weeks, 34m (110') of tunnel was successfully stabilized with 21,800kg (48,000 lb.) of PU.

Most PU materials are extremely water-sensitive, creating a weak, low-density material in wet conditions. Micon's RokLok 70 is a hydrophilac, controlled expansion, two-component PU material that draws little moisture into the reaction process, creating a stronger, denser end product. Typical use was to inject Roklok-70 through 2 to 6, 4.6-6m (15-20') long, 12mm (½") diameter grout pipes. The loose nature of the ground where PU was required resulted in many of the pipes simply being pushed by hand or with the assistance of a 5.4 kg (12 lb) hammer. Schedule 80 pipe was used due to grout pressures up to 100 bar (1500 psi).

Shortly after entering Reach 1A (granodiorite), blocky ground was encountered. As the TBM drives continued, ground conditions deteriorated in shear and fault zones with rock collapsing ahead of the face of the TBM's resulting in ravelling and voids. Again, PU grout was successfully used as a void filler/rock binder to create a 'glued arch' above the TBM and permit mining to continue, albeit only in short advances until the cutterhead mined beyond the grouted zone or into more competent ground. This situation recurred several times in Reach 1 with another 78,500kg (173,000 lb) of polyurethane grout used in this manner. The application of polyurethane contributed greatly to successful TBM mining through the varying geology beneath the Santa Monica Mountains.

Stage Grouting from Special Seismic Section

At the south end of the drive, an oversized tunnel or special seismic section was incorporated into the tunnels to cross the active Hollywood Fault. This would be the terminus of the TBM tunnels that were being mined from the north and was originally to be mined by drill and blast methods. However, a horizontal borehole confirmed the ground to consist of intensely sheared, brecciated and decomposed granodiorite and the design was altered to include lattice girders and mining by NATM using roadheaders. It became clear as excavation through the seismic section proceeded that ground conditions were not improving as expected. Poor ground was still present at the end of the 91m (300') excavation and ground conditions were thought unsuitable for excavation by TBM. Options considered were:

• Continue excavation with the seismic section design or with a smaller diameter using steel ribs until ground conditions improved.
• Grout to improve the ground conditions for excavation by TBM.

Another horizontal 90m (300') core hole was taken which showed that the sheared rock continued for at least another 85m (280'). Grading tests on the worst of the core material revealed that approximately 85% of the material had grain size greater than 0.075 mm (200 sieve) and thus the fault gouge material was believed suitable for permeation grouting with microfine cement. Cost and schedule analyses confirmed the decision to improve the ground for TBM excavation despite considerable risk as some parties felt that treatment alone would be insufficient to enable TBM progress through the sheared ground.

Due to the excavation sequence in the seismic section, grouting was carried out in the AL top heading first followed by the AR top heading with bench grouting following in a similar manner. Initial plans called for use of sleeve port grout pipes (SPGP) and a systematic grout injection sequence. However, problems with the installation of the SPGP's in the broken ground led to a stage grouting approach.

In total, eleven 100mm (4") diameter cased horizontal holes were drilled in each tunnel using a track mounted Klemm duplex drill rig. Holes were located around the periphery of the approaching TBM tunnel (see Figures 6 and 7) and were initially drilled to 11m (35') depth. These holes were grouted to 7 bar (100 psi) pressure refusal with microfine cement to create a "bulkhead" against which higher grouting pressures could be applied. All holes were re-drilled in stages to depths of 30, 45, 60 and 85m (100, 150, 200, and 280'). Each stage length was grouted to a refusal pressure of 10-15 bar (150-200 psi) with 4 holes being grouted simultaneously.

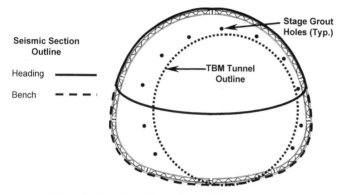

Figure 6: Elevation of Stage Grout Hole Pattern at end of Seismic Section.

Figure 7: Hole Drilling for Stage Grouting.

Grouting was completed well in advance of TBM arrival and grout takes were surprisingly high with a total of 165,300 kg (364,500 lb) of microfine cement pumped into the ground. The first indication of grouting operation success was the reduction in water inflow from 250 l/min (70 gpm) to less than 70 l/min (20 gpm) in a 75m (250') trial hole drilled in the center of the grouted zone.

The success of the treatment was proven by TBM excavation progress in the totally sheared material of Reach 1C. At approximately 170m (560') from the seismic section, progress immediately slowed and polyurethane grouting was required in both tunnels in order to proceed. However, once the stage grouted area was reached, the ground improvement was such that excavation rates increased significantly (see Table 3) with a minimum saving of 18 days of TBM operation time.

Distance from Seismic Section:	TBM Progress Rate
Untreated: 170 to 85m (560 to 280')	6.4 m/day (21 ft/day)
Stage Grouted: 85 to 0m (280 to 0')	18.9 m/day (62 ft/day)

Table 3: Increased TBM Rates due to Stage Grouting

Conclusion

The AR TBM successfully holed through to the seismic section on October 22, 1997 shortly followed by the AL machine on November 7, 1997. The successful completion of these tunnels was made possible by the implementation of various grouting methods for ground improvement, water inflow control, environmental issues and settlement limitations and the flexibility of the owner, designer, contractor and construction manager.

References

Los Angeles County MTA. (1994). "Contract C0311 Tunnel Line Section – Sta 630+00 to Universal City Geotechnical Design Summary Report," Los Angeles County MTA.

Kramer, G.J.E. and Albino, J. (1997). "Los Angeles Metro - TBM Starter Tunnels - Mixed Face Conditions," *Proceedings ASCE* Construction Congress V, Minneapolis, Minnesota (pp. 241-250).

Taylor, G.E., Cavey, J.K. and Roach, M.F. (1997). "Horizontal Grouting Beneath the Hollywood Freeway. Los Angeles MetroRail C-311," *Proceedings* Rapid Excavation and Tunneling Conference, Las Vegas, Nevada (pp. 73-81).

Acknowledgments

The Authors wish to acknowledge the Los Angeles County Metropolitan Transportation Authority for allowing us to present the information contained in this paper. In particular, we wish to acknowledge Mr. Fred Smith, Construction Site Manager for LACMTA for his review and insightful comments during the preparation of the paper.

JET GROUTING FOR THE 63rd STREET TUNNEL

F. Pepe Jr.[1], M. ASCE, G. A. Munfakh[1], M. ASCE, and Yves St-Amour[2]

ABSTRACT

Jet grouting was used as part of the excavation support, underpinning and groundwater cutoff systems used for the construction of the 63rd Street Tunnel Connection Project in New York City. Jet grouted walls were installed below the groundwater level and to depths of up to 34.5 meters in difficult glacial deposits. Innovative construction techniques and equipment were used to install the wall elements in low overhead conditions and through the existing subway structure while maintaining continuous subway service throughout construction. Extensive verification and testing programs were performed during both the design and construction phases of the project. A circular test shaft of jet grouted columns was installed during the design phase to determine wall installation parameters and verify the suitability of the jet grouting technique for the difficult glacial deposits present at the site. Quality control testing during construction included continuous monitoring of the jet grout parameters using a real time data collection system, vertical measurements at each grout column, coring, packer permeability testing, and a pump test at the completed installation. This paper describes the innovative aspects of the design and construction of the jet grouting works, the verification and testing programs used, and the performance of the jet-grouted walls installed for the project.

INTRODUCTION

The 63rd Street Line Connection Project is part of the New York City Transit (NYCT) master plan to extend subway service and alleviate congestion on the existing subway lines connecting the Manhattan and Queens Boroughs in New York City. The project, when completed, will connect the existing 63rd Street and Queens Boulevard transit lines between 29th Street and Northern Boulevard in

---

[1] Senior Professional Associate and Vice President, respectively, Parsons Brinckerhoff Quade & Douglas, Inc., One Penn Plaza, New York, New York 10119

[2] Technical Director, Pacchiosi North America Inc., 3120 Frechette Street, St-Jean-Baptiste de Rouville, (Quebec) Canada J0L 2B0

Long Island City.   Figure 1 shows the project location and the complex
arrangement of roadway and transit lines existing at the site.

The new subway connection will include construction of approximately 480 lineal
meters of new subway tunnel using both cut-and cover and mined tunnel
techniques.   The major portion of the new tunnel will be constructed beneath
active subway lines, which will remain in service throughout the work.   The new
connection will be made by tunneling under the existing NYCT subway structure
near 49th Road and Northern Boulevard and then gradually rising to meet and
penetrate the existing subway structure's invert near Honeywell Street as shown in
Figure 1.   The construction will include cuts of between 9 and 24 meters as well as
dewatering of up to 15 meters.   The work will be completed while maintaining
continuous train service on both the underground Queens Boulevard line and the
elevated Astoria line (Figure 1) as well as maintaining vehicular traffic on the
surface roads above.

Due to the complex nature of the work, the project was packaged into three major
construction contracts.   Each contract had specific challenges that required
innovative designs and special construction techniques.   This paper addresses the
jet grouting work done as part of the design phase test program and Construction
Contract C-20201 that included the majority of jet grouting performed on the
project.

PROJECT DESCRIPTION

The existing Queens Boulevard subway line which is also known as the IND line,
runs beneath Northern Boulevard through a five-tracked jacked arch box structure
with an invert approximately 7.5 meters below the street grade.   The existing
tunnel, which carries more passengers into Manhattan than any other subway line
in the NYCT system, was constructed in 1923 using the cut-and-cover method of
construction.   The new connection in Contract C-20201 consists of two new
subway tracks that run beneath and penetrate into the existing structure as shown
in Figure 2. A local by-pass track will also be added at the existing invert level as
part of the connection project.

A key element of the construction of the new connection tunnels is that the work
must be performed in a watertight enclosure formed by cutoff walls.   The
watertight enclosure is needed to permit dewatering inside, where the tunnels will
be built, while minimizing drawdown of the water table outside of the enclosure.
Control of the groundwater table is required to prevent migration of an adjacent
PCB-contaminated plume, as well as settlement of peat and organic silt layers that
underlie structures at the site.

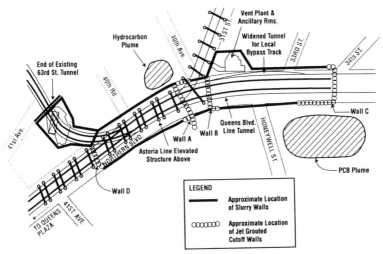

Figure 1. Approximate Location of Jet Grouted Walls

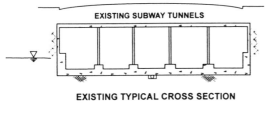

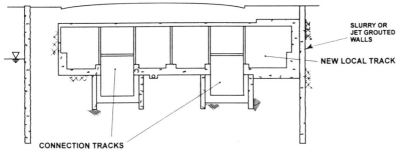

Figure 2. New Subway Connection Construction

The cutoff and excavation support walls consist of a combination of slurry diaphragm walls and jet grouted walls as shown in Figure 1. Four jet grouted walls, designated as walls A, B, C, D were constructed for the entire project. Pacchiosi North America, Inc. constructed walls A, B and C while Wall D was constructed by a joint venture of Hayward Baker, Inc. and the Nicholson Construction Company. The jet-grouted walls supplement the slurry walls running parallel to the existing subway line and form transverse walls under the subway. The jet grouting was used in areas where conventional slurry walls could not be constructed due to the presence of existing subsurface and surface structures and utilities.

A total length of 155 meters of jet grouted walls consisting of about 550 individual columns were installed for walls A, B, and C (Figure 1). The columns extended to depths of about 34.5 meters from the ground surface. Each wall was constructed of three rows of overlapping columns spaced on an average spacing of 840 mm centers for a total wall thickness of between 2.1 and 2.4 meters. A minimum overlapping thickness of 300 mm was required between adjacent columns.

A key requirement of the jet grouting program implemented on this project was the extensive quality control and assurance testing performed during both the design and construction phases of the project. During the design phase, a jet grouting test program which included extensive instrumentation was developed and conducted to provide data on the performance of jet grouting in the various deposits that occur at the project site. The results of the design phase test program were used to evaluate the suitability of jet grouting at the project site and develop the design and acceptance criteria for production work. The quality control and assurance testing programs conducted during the production grouting were developed to monitor construction and assess if the key design criteria were achieved. A test program was performed at the beginning of construction to select the installation parameters to be used in the production grouting.

SUBSURFACE CONDITIONS

The project site is underlain by metamorphic bedrock, which is covered by glacial, interglacial, and post glacial deposits. The glacial deposits can generally be divided into three main groups: mixed glacial deposits, glacial till, and outwash/reworked till deposits. Stratification of these deposits is complex and significant variations in the thickness and distribution of the individual units is common (Figure 3). This heterogeneity is typical of glacial deposits found at the rear of terminal moraines.

A generalized subsurface profile at the project site (Figure 3) consists of:

- Miscellaneous fill (Stratum 1) consisting of a heterogeneous mixture of course to fine sand, silt, gravel, brick fragments, wood, metal, and other building rubble. The deposit is generally loose to dense in consistency.

- Mixed glacial deposits (Strata 2, 3, & 4) varying from loose-to-medium dense coarse to fine sand, to soft-to-stiff varved silt and clay and very fine sand. This group contains glacial, lacustrine and glaciofluvial deposits.

- Outwash/reworked glacial till deposits (Stratum 5A) consisting of dense to very dense, well to poorly graded coarse to fine sand. Boulders and cobble size particles are also commonly encountered in these deposits.

- Glacial till deposits (Stratum 5) generally interlayered with the reworked till and outwash deposits and consisting of heterogeneous mixture of sand, silt and gravel with and without a cohesive binder. The till materials are generally dense to very dense, contain numerous boulders and cobbles.

- Bedrock (Strata 6, 7A, and 7B) ranging from totally decomposed to slightly fractured to sound granite gneiss to schistose gneiss with occasional quartz pegmatite veins.

- Peat and organic silt (Stratum 8) consisting of soft to medium stiff compressible peat and organic silt. This deposit is encountered randomly under the site and underlies the fill deposits as well as the existing subway box structure in local areas.

DESIGN PHASE JET GROUTING TEST PROGRAM

This design phase test program (Figure 4) was developed to address the following issues: (1) jet grout column strength, (2) jet grout wall continuity and encapsulation of boulders, (3) jet grout column diameter, (4) jet grout wall permeability, (5) potential for ground displacements during jet grouting, and (6) pressure dissipation in the ground adjacent to the jet grout monitor.

To address these issues, a three-phase test program was implemented as follows:

- Phase I - Installation of overlapping jet grout columns to construct a circular test cell 9 meters deep, with a 3-meter deep jet-grouted bottom plug.

- Phase II - Diamond core drilling, in-situ packer permeability testing of columns, and laboratory strength testing of jet grouted soil samples.

- Phase III - Excavation of the test cell, dewatering inside of the cell, measurement of seepage into and drawdown outside of the excavated cell, and observation and mapping of the exposed grouted mass.

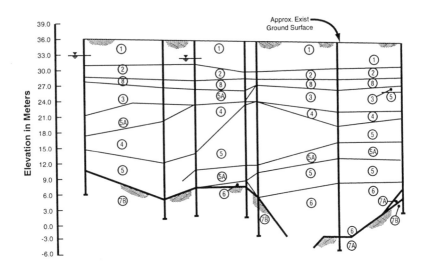

Figure 3. Typical Subsurface Profile

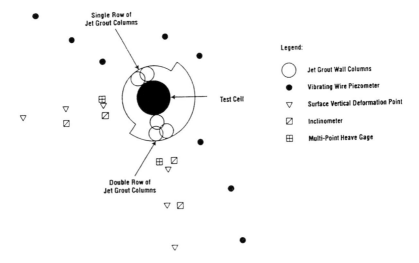

Figure 4 - Jet Grout Test Cell and Instrumentation Layout Plan

Figure 4 presents the column arrangement for the test cell as well as the layout of instrumentation installed in the vicinity of the cell. The inside diameter of the cell was 2.4 meters. Two different configurations of jet grout columns were used to form the cell walls. One half of the cell was formed by a single row of columns installed at 825 mm centers and the other half by a double row with the outermost columns installed at 875 mm centers. The triple fluid jet grouting system was needed to install all of the jet grout columns. The instrumentation program included inclinometers, piezometers, and surface and deep settlement points. Hayward Baker, Inc. performed the jet grouting for the design phase test program.

A detailed discussion of the results of this test program is included in two publications on the subject (New York City Transit Consultants, 1994, and Flanagan and Pepe, 1997). The main conclusion of the test program was that jet grouting was applicable for the 63rd Street Connection Project but that modifications to the design criteria and specifications were needed for the production grouting. The key requirements included in the bidding documents for the production grouting included:

- The walls should consist of at least three rows of overlapping columns for a minimum wall thickness of 2.1 meters.
- The walls should have a maximum permeability of 1×10^{-6} cm/sec and an unconfined compressive strength of 5.5 MPa at 28 days.
- The quality control/assurance program for the production grouting should include: a) pre-grouting exploratory borings, b) intensive coring and packer permeability testing of the jet grouted walls, c) laboratory strength tests of cored samples, d) real time recording of operating parameters during grouting, e) pre-production field trial using the Contractor's proposed methods and procedures, f) displacement monitoring of the active subway lines and adjacent structures, g) checking of the verticality of the jet grouted columns, and h) performance of pumping tests for the entire enclosed cutoff wall system with measurement of internal and external drawdowns.

PRODUCTION JET GROUTING CHALLENGES

The production jet grout work offered a great number of challenges that required close cooperation between the owner, engineer, general contractor and specialist jet grouting subcontractor. These challenges encompassed both the application of the jet grouting technique, as well as the modification of the jet grouting equipment to satisfy the site conditions. Specifically, the jet grouting work required addressing the following construction issues:

1. Two-thirds of the jet grout walls were located directly under the existing invert of the IND subway tunnel running underneath Northern Blvd. and the jet grouting had to be accomplished without interruption of subway service to and from Manhattan. This meant that jet grouting for a large segment of the work

could only be accomplished on weekends when selected tracks (maximum two at any one time) could be shut down for a 54 hour period.

2. Since subway service was continuous during construction, it was critical that the structural integrity of the IND structure be maintained at all times; that no water be allowed to seep into the existing subway; and perhaps of most importance, that any potential for heave, settlement or loss of support of the existing subway structure be avoided.

3. Approximately one-third of the jet grout walls were located outside of the IND subway box structure, generally in locations where access was difficult, and where numerous existing underground utilities existed and could not be removed.

4. All the jet grouting work was executed either directly below or immediately adjacent to Northern Blvd. which is a major arterial street in Queens, New York, and is one of the two main access streets to the 59th Street Bridge linking the Boroughs of Queens and Manhattan. Therefore, all jet grouting work had to be accomplished during temporary lane shutdowns while traffic on Northern Blvd. remained open in both directions at all times.

JET GROUTING SYSTEM

Upon evaluation of the above challenges, as well as site geology, and construction and access issues, the triple fluid method of jet grouting (Pacchiosi PS-3) was selected to construct walls A, B and C. Among the key reasons for selecting the PS-3 method were: a) the heterogeneous nature of the soil and rock geology, b) the need to minimize the potential for heave of the subway track structure during construction, and c) the requirement to minimize the number of columns necessary to achieve a continuous cutoff wall.

The PS-3 triple fluid system begins with the drilling of a small diameter hole to the bottom elevation of the jet grout column using standard rotary or percussion drilling methods. Once the bottom depth is reached, three high pressure fluid jets (air, water, grout) are activated to cut and form the jet grout column. The PS-3 jet grout system uses the erosive action of a high pressure water jet shrouded in an envelope of compressed air to cut and partially excavate the in-situ soils. As the column is being formed by the cutting action of the water jet, the soils that are displaced to the surface are replaced by cement grout introduced by a separate nozzle located at the bottom of the drill rod stem. The cement grout mixes with the eroded soil to form the jet grout column.

By slowly withdrawing the continuously rotating drill string it is possible, in certain soil conditions, to form jet grout columns exceeding 2 meters in diameter. In the case of C-20201, after analysis of the soil conditions, the Contractor planned to form jet grout columns approximately 1.5 meters in diameter.

EQUIPMENT CONSIDERATIONS

Drilling Equipment
Because of the variability and complexity of the subsurface conditions, an early decision was made to use a drilling system which could accommodate a wide variety of drilling techniques, including standard rotary drilling, as well as drilling with down-the-hole hammers, including both air and water hammers.

To deal with low overhead conditions (less than 4.5 meters of clearance beneath the elevated Astoria train tracks) at the south end of the project site, a low overhead drill rig was specially manufactured to include a carousel rack feed system. This system permitted rapid adding and removing of drill rods to the drill string

For the unobstructed areas of the site where no overhead restrictions existed, a standard drill rig equipped with a 30 meter mast was used. This minimized the need for manually adding or removing sections of drill rods, and allowed most of the jet grout columns to be drilled and grouted in a single continuous operation.

Pumping/Mixing Equipment
At the heart of any jet grouting operation is the grout mixing and pumping system used to deliver the grout at very high pressures to the point of injection. For the 63rd Street project, two complete jet grout pumping and mixing plants were mobilized for the duration of the work. Each mixing unit consisted of a fully containerized, automated grout mixing plant with cement storage capacity of two 54 tonne cement silos per plant. Each pumping unit consisted of a fully containerized, hydraulic high pressure pump plant equipped with dual pistons capable of simultaneously delivering both grout and water at very high pressures required for the PS-3 triple fluid application.

Data Collection and Monitoring System
To control the jet grouting process, the mixing and pumping plant as well as the jet grout drill rigs came with a real time jet grout data collection and monitoring system (PRS-3). Sensors placed on the mixing, pumping and drilling equipment collected real time data that allowed for constant monitoring of the key parameters of the jet grout operation.

FIELD PROCEDURES

Two basic groups of jet grout columns were installed: one below the NYCT IND subway box and the other outside that box. For either group, site conditions, such as utilities or obstructions, required incremental adjustments during the course of the work to the basic jet grouting procedures. Approximately two-thirds of the jet grout columns were drilled through the IND box structure. The remaining columns were located outside that box. These columns were installed at existing buildings, utilities, a subway entrance structure, or were connection points to the sewer siphon at the south end of the project site.

Installation through the IND Box

The procedure used for columns installed through the IND box, illustrated in Figures 5 and 6, was as follows:

1. For each wall, the general contractor first excavated, from the street level to the top of the IND roof structure, a six foot wide by eight foot deep trench. Road plates were then placed over the trench to permit vehicular traffic to cross.

2. A 150 mm hole was cored through the roof of the IND box from within the trench at the location of each jet grout column for about two-thirds of the invert concrete thickness. A steel sleeve was then installed into the invert hole and cemented in place.

3. To penetrate the subway invert, a drill casing was lowered from the street level through the roof and screwed into the steel sleeve inserted during the coring operation. Once the drill casing was in place, the drill rod was lowered through the casing to the tunnel invert. A special drill chuck and jet grout return flow collection fitting was then set into place at the top of the drill casing. This fitting permitted the return flow to be sent to a small collection tank and then to be pumped to the return flow collection pond. In this manner no fluid of any type entered into the IND box.

Jet Grout Column Formation

Once the casing and return flow collection fittings were in place, drilling would commence through the invert concrete into the soil and rock beneath the IND box. Drilling operations using a polymer drilling fluid then would drill down a 110 mm diameter hole to the bottom elevation of the jet grout column. The drilling fluid was used to reduce friction at the drill bit, stabilize the side walls of the drill hole from collapse, and provide a fluid transport medium for the removal of cuttings to the surface.

After the drilling operation was completed, the vertical hole deviation was measured at full depth using a specially developed inclinometer probe lowered inside the center tube of the triple tube rod. Verticality measurement was then made to verify that the vertical deviation at any depth along the jet grout column would be within acceptable tolerances to ensure effective continuity of the jet grout cutoff wall. If the measured deviation was more than 2%, either the hole would be abandoned after being backfilled with grout, or additional jet grout columns would be added to close the potential gap in the wall. When drilling deviation was within acceptable tolerances, which was the case for more than 99% of the columns, the grouting operation would begin.

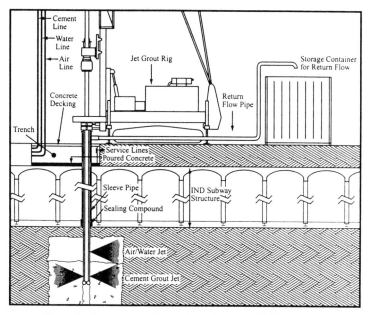

Figure 5 - Jet Grouting Proceedure Below the IND Structure

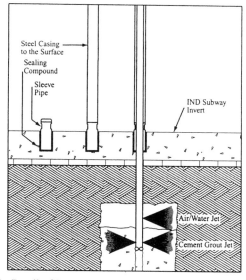

Figure 6 - Detail of Invert Penetration through the IND Subway Invert.

The column formation process was then initiated by activating the air, water, and cement grout jet. The jet grouting injection parameters used for the project are presented in Table 1. Return flow of the injection was collected at the surface through the outer protective steel casing. Once the injection monitor reached the invert slab, the high pressure jets were deactivated, the outer casing was removed to the surface, and the invert sleeve was sealed with a cap.

After the grout was set, the sleeve cap was removed to check for any subsidence at the top of each column. In the event that a column had experienced subsidence, which was the case for approximately 20% of the columns, the top of the column was grouted with a low pressure cement grout pump, using a 2:1 cement to water ratio grout.

Table 1 - Jet Grout Parameters for 63rd Street Project

Soil Formations and Depths	Injection Parameter	Parameter Value
Upper Fill, Organic Layers	Column Withdrawal Rate (minutes/meter)	6.25
and	Grout Specific Gravity	1.65
Clayey or Silty Layers	Water/Cement Ratio	0.72
	Cement Grout Pressure (MPa)	13
Depths - 2.4 to 12 meters	Cement Grout Flow Rate (l/min)	130
from ground surface	Water Pressure (MPa)	45
	Water Flow Rate (l/min)	100
	Air Pressure (Mpa)	175
	Air Flow Rate (l/min)	60
	Rotational Speed (RPM)	5
Lower Dense to Very Dense Till	Column Withdrawal Rate (minutes/meter)	10.5
	Grout Specific Gravity	1.65
Depths from 12 meters to	Water/Cement Ratio	0.72
bottom of column (max.	Cement Grout Pressure (MPa)	18
depth 34.5 meters from	Cement Grout Flow (l/min)	150
ground surface)	Water Pressure (MPa)	45
	Water Flow (l/min)	100
	Air Pressure (MPa)	175
	Air Flow (l/min)	60
	Rotational Speed (RPM)	5

Minimizing Heave Under The IND Structure

Approximately one-third of the way into the completion of jet grout Wall B, monitoring of the IND subway tracks revealed heaving of the subway tunnel structure of up to 35 mm. The timely monitoring and response by the design and construction team prevented damage to the IND structure. However, additional measures to prevent further movements were implemented, and included the following:

- Drilling of a 150 mm diameter hole through the entire thickness of the tunnel invert.

- The drill bit diameter was increased from 100 to 125 mm in the soil to provide a larger annulus around the drill rods for channeling the return flow to the surface.

- Thicker polymer slurry was used during the drilling operation to remove more cuttings to the surface, thereby providing a cleaner and more stable hole for the column injection operations.

- Relief holes were installed through the tunnel invert in the vicinity of each injection hole to prevent the build-up of pressures immediately below the invert.

- Grout pressures were reduced for the top 1.5 meters of the column immediately below the invert.

- Continuous monitoring of the tracks was implemented during the remainder of the jet grouting work. The jet grout work was to have been suspended should the monitoring disclose movements of more than 3 mm.

Following the field implementation of the heave remediation measures, additional movement of the IND structure during jet grouting was minimal and well within expected tolerances. Less than a 1.5 mm of movement was recorded during the duration of the remaining jet grout work which included the installation of the remaining two-thirds of Wall B and all of Wall A.

QUALITY CONTROL TESTING

A significant factor in the successful application of jet grouting on the project was the incorporation of an adequate quality control (QC) and verification testing program. The properly planned and executed program resulted in early identification of potential problems and allowed the contractor to make proper modifications to deal with these problems. The QC program was implemented during all phases of the project.

Basically, the QC procedures were implemented in four phases: 1) design phase, 2) pre-production phase, 3) production phase, and 4) post-production phase. The owner, engineer and contractor were involved in implementing these procedures.

Design Phase QC Work

During the design phase, geotechnical investigations and analyses were performed to define the subsurface conditions and establish the overall requirements for jet grouting. Specific ground conditions such as boulders and peat that might affect the performance of jet grouting were identified and their impact evaluated. The design phase jet grouting test program was also performed to collect site specific data for final design. Other factors which might influence the quality of the jet-grouted mass were considered in the design and preparation of construction specifications for the project.

Pre-Production QC Work

As part of the construction contract, a pre-jet grouting exploratory program was performed to confirm the soil and rock conditions along the walls. The pre-jet grouting exploratory holes were taken every 3 meters along the wall alignment. Exploratory holes included soil sampling, rock coring and rock packer permeability testing.

A pre-production test program was implemented after completion of the pre-jet grouting exploratory borings. Two sets of test columns were installed prior to the start of production grotuing: one for Wall C and one for walls A and B (Figure 1). The test columns were installed using actual production parameters and equipment, and then tested to determine the suitability of the jet grout columns and establish the required injection parameters and procedures. Full length cores were taken from the completed columns for strength testing, and packer permeability tests were performed within the bore hole.

Production QC Work

The production QC work consisted of real time data collection and monitoring of the entire grouting operation, plus monitoring of the verticality of the jet grouted column. A real time data collection system monitored the mixing, pumping, drilling and grouting works for each completed column. The collected data was downloaded to an on-site computer program and condensed into a series of graphs, which plotted the injection parameters in relation to time and depth. This permitted the engineer and the contractor to have a complete quantitative documentation for each column, and allowed verification that each jet grout element had been constructed in the prescribed manner.

The verticality of the completed column was measured by a small-diameter inclinometer lowered through the center hole of the injection rod to the bottom elevation of the column. The inclinometer probe was then withdrawn and readings were taken every 3 meters to measure both the X and Y deviation from the center axis of the column.

Table 2 lists the parameters recorded by the PRS-3 system.

Table 2
Jet Grouting Parameters Recorded by PRS-3 System

Phase	Recorded Parameter
Drilling Phase	Depth Drilling Rate Pressure or Drill Tools Rotary Pressure Drilling and Pressure Drilling Fluid Flow Rate
Grouting Phase	Depth Water Pressure Water Flow Rate Grout Specific Density Grout Pressure Grout Flow Rate Air Pressure Air Flow Rate

Post-Production QC Work

The post-production QC work included laboratory strength testing of core samples taken from the hardened columns, in-situ packer permeability tests performed within the cored columns, and a full-scale pumping test performed in the vicinity of the completed cut-off wall.

TEST RESULTS

A total of 53 unconfined compressive strength tests were performed on the core samples. The test results met or exceeded the specified minimum strength of 5.5 MPa in 97% of the tests. The unconfined compressive strength of the jet-grouted soil ranged from 4.3 to 33.1 MPa with an average of 12.3 MPa. No relationship between strength and soil type was observed due to the heterogeneous nature of the soil conditions at the site.

A total of 151 packer permeability tests were conducted for the project. The permeability ranged from 4×10^{-5} to less than 1×10^{-7} cm/sec with an average of 2.9×10^{-6} cm/sec. Approximately 84% of the measured values satisfied the required specification of 1×10^{-6} cm/sec. The results of these tests, observations during the pump test, and excavation indicated that the tight overlapping triple rows of jet grouted columns are generally effective. However, the results of the packer tests required some remediation to address local areas of high packer test permeability.

REMEDIATION WORK

In areas where the QC results showed deviation from the requirements of the contract specifications, remediation works were implemented. Remediation in these cases consisted of chemical grout injections using sodium silicate grouts.

CONCLUSIONS

The following conclusions can be drawn from this case study:

- Jet grouting is an effective method of constructing in-situ walls for excavation support or as cut-off barriers.

- The jet grouted wall can be constructed in areas of low headroom where conventional excavation-support systems, such as slurry diaphragm walls and soldier pile and lagging, cannot be used.

- An added advantage of jet grouting is the ability to construct the wall through boulders and obstructions. The triple fluid system is the most suitable jet grouting method for applications similar to that of the 63rd Street Tunnel Connection project.

- The unconfined compressive strength of the jet-grouted soil ranged from 4.3 to 33.1 MPa, with 97 percent of the test results exceeding the minimum specified strength of 5.5 Mpa. No relationship could be established between the measured strength and the soil type due to the heterogeneous nature of the soil conditions at the site.

- The in-situ permeability of the jet-grouted column ranged from 4×10^{-5} to less than 1×10^{-7} cm/sec. The use of three rows of overlapping jet-grouted columns is effective in creating a cut-off wall.

- Quality control is a significant factor in the successful application of jet grouting. A four phase QC program is recommended. A real-time data collection and monitoring system can be used to control the injection parameters and assess the need for remedial measures.

- Column verticality is an important factor that influences the effectiveness of the cut-off wall. Special inclinometers can be used to verify the column verticality.

- A field trial is a must for each jet grouting project. The pre-production trial assists in evaluating the effectiveness of the equipment to be used and selection of the appropriate injection parameters.

- Pre-qualification of jet grouting contractors is necessary. Detailed prequalification questionnaires should be included in the bidding documents.

- Close cooperation between the owner, engineer, and contractor is critical for the successful implementation of a jet grouting project.

REFERENCES

Flanagan, R., Pepe, F. (1997), "Jet Grouting for a Groundwater Cutoff Wall in Difficult Glacial Soil Deposits", International Containment Technology, Conference and Exhibition, Florida, pp. 248-255.

New York City Transit Consultants (1994), "Jet Grouting Test Program Report", final report to New York City Transit Authority.

GROUTING TO CONTROL COAL MINE SUBSIDENCE

Donald E. Stump Jr. P.E., Member ASCE[1]

Abstract

Abandoned underground coal mines are located throughout the coal fields in the United States. The mines are anywhere from a few meters to several hundred meters below the ground surface. In many of these mines coal pillars and/or wooden supports were left in the mine to keep the roof from collapsing during operations. After abandonment these roof supports deteriorate and eventually collapse causing subsidence. To protect the public from subsidence and other adverse effects of coal mining an Abandoned Mine Lands Reclamation Fund was established under Public Law 95-87. This fund is used to finance subsidence control projects. Grout is frequently used to stabilize areas in the mines by filling the mine voids and providing the support needed to minimize future subsidence. However, the placement of the grout may require unique equipment or customized grout materials since abandoned mines are generally inaccessible, the areas requiring stabilization are beneath surface structures, the voids may be flooded, and methane gas may be present. In addition to special considerations in grout placement the areas requiring roof support may be extensive and filling the voids in the entire mine are cost prohibitive. Special grout placement techniques have been evaluated including controlled grout columns or synthetic bags filled with grout and chemical additives to control grout set-up.

---

[1]Civil Engineer, United States Department of the Interior, Office of Surface Mining, 3 Parkway Center, Pittsburgh, PA 15220

INTRODUCTION

Coal is located throughout the United States and has been mined since the 1600's (Office of Technology Assessment 1979). Underground coal mines create voids where the coal is removed. During mining a percentage of the coal or cut timbers were left in place to support the mine roof. If this support fails the mine roof collapses and the rock and soil above the mine will sag or fracture causing the ground surface to subside.

In areas where coal mine subsidence needs to be controlled additional support of the mine roof has to be constructed. Abandoned underground mines are usually inaccessible and the construction of these roof supports can to be done by injecting material through boreholes. Traditionally loose granular material or grout has been used to provide roof support. However, the cost of using these traditional techniques may be prohibitive depending on the size of the area that requires roof support.

To minimize the amount of material required to provide roof support a variety of grout mixes and grouting techniques have been developed including, controlled grout columns, synthetic bags filled with grout, and chemical additives to control grout set-up.

COAL MINE SUBSIDENCE

Coal has been found in almost every state in the United States and nearly one-eighth of the country's land lies over coal beds (Office of Technology Assessment 1979). In early years, coal was mined in largely agricultural areas remote from centers of population and as a result, surface subsidence was not a major problem. However, in recent decades expanding suburbs have resulted in the construction of many buildings over abandoned underground mines with relatively shallow cover (Gray and Meyers 1970).

Subsidence is the sagging or falling of a mine roof because of mine roof collapse, mine floor movement or coal pillar crushing (Craft 1990). The ground surface expression of subsidence are commonly categorized as either sinkholes or troughs. Sinkholes are depressions or holes in the ground surface (Figure 1) and troughs are shallow, often broad, dish-shaped depressions (Figure 2).

Subsidence may occur concurrently with mining or may be delayed until long after mining is completed. It is estimated that over 2.8 million hectares of land in the United States are underlain by abandoned coal mines. Approximately 0.8 million have already been affected by subsidence and the remaining 2 million hectares are susceptible to future subsidence (Office of Surface Mining 1982).

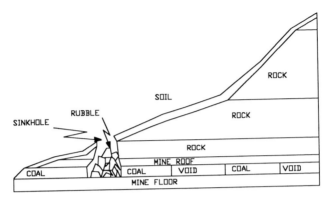

Figure 1 : Sinkhole

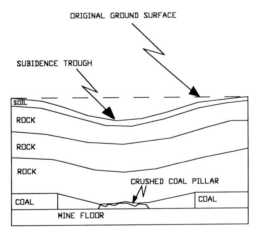

Figure 2 : Trough

ABANDONED MINE LANDS PROGRAM

On August 3, 1977 the Surface Mining Control and Reclamation Act (SMCRA) was signed by President Jimmy Carter. SMCRA provides authority for regulating coal mining and reclamation on public and private lands,. SMCRA is administered by the Office of Surface Mining (OSM).

One of the major programs created by SMCRA is the reclamation program for abandoned mine lands, funded by fees that coal mine operators pay for coal mined. The fees vary depending on the type of mining (surface or underground) and the type of coal removed (lignite, bituminous, subbituminous, or anthracite). These fees are distributed to the states where abandoned coal mine reclamation work needs to be done. Moneys from the fund are also used to fund the emergency program. Emergency projects are those involving abandoned mine lands that present a danger to public health, safety, or general welfare and which require immediate action (Office of Surface Mining 1998).

SUBSIDENCE CONTROL TECHNIQUES

Backfilling

Backfilling abandoned mine voids is a common method of stabilization used to prevent or abate subsidence and protect surface structures. Backfilling is done to fill the mine voids with granular material and may be placed in a controlled fashion if the mine is accessible or may have to be done remotely through boreholes. The material may be injected with air, gravity fed or mixed with water to form a slurry.

A problem with backfilling is the fact that granular backfill, even if compacted, is somewhat compressible when subjected to the weight of collapsing overburden. Therefore, even a completely backfilled mine may continue to subside.

Grouting

Grouting is another method for stabilizing abandoned mines. Grout is typically a cement/sand/water mixture which after injection, gains strength and becomes relatively incompressible. Grouting is normally applied to stabilize areas beneath specific structures rather than supporting relatively large open areas.

The areas grouted to provide support are designed to have a compressive strength equivalent to that of coal pillars. The average compressive strength of bituminous coal pillars is assumed to be 130 kPa. This is based on testing done on small pillars of coal from the Pittsburgh seam (Ackenheil and Dougherty 1970).

The spacing of the support columns is based on the depth of the mine from the

ground surface, the thickness of competent overlying rock, the location and
physical arrangement of any existing coal pillars, and the bearing capacity of the
mine floor. Another key component in the design of the support column is to
maximize the area of roof contact with the column. In areas where the mine roof
slopes large area contact with the roof is difficult.

Grouting is equivalent to low volume pumped-slurry injection in that a closed
injection system is used and the fill material is pumped from a central mixing
location to several injection holes.   If the grout is placed under only the static head
imposed by the column of grout, the method is known as gravity grouting (Figure
3). Where considerable subsurface caving has occurred, penetration into joints in
the overlying strata is generally desired and grout is often injected with controlled
pressure.

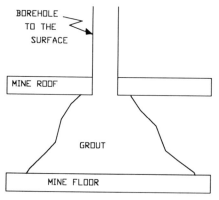

Figure 3 : Grout Column

In grouting continuous mine voids where little caving has occurred, a relatively thick grout mixture is generally injected into gravel barriers at mine level to form a wall around the area. A thick mixture is used so the barrier will maintain a steep (one vertical to two horizontal) angle of repose, reducing the loss of material outside the area being stabilized. Once an impermeable wall is formed, the remainder of the area is filled with more fluid grout. Injection is often conducted in stages, with either thin grout or expansive grout being used in the final stages to achieve good roof contact. Borehole spacing is typically 10 to 12 meters on center, with 15 cm diameter holes.

Grout mixtures vary depending upon the requirements of the individual project. For most applications, Portland cement and sand and fly ash mixtures are used because of low cost and the reliability of strength gain. Mixtures of one part cement to 3 to 9 parts sand or fly ash are typical. Where increased penetration is required or where quick sets are desirable, chemical grout mixes could be used, but are often several times more expensive than cement grout mixes.

Controlled Grout Columns

When the volume of grout required to fill an underground mine void becomes to large and costly or control of the placement of grout is difficult an alternate grout placement method should be considered. One alternate method that requires significantly less grout has been used are cylindrical columns of grout constructed by filling synthetic fabric bags (Michael, Lees, Crandall, and Craft 1987).

This method is appropriate where the mine roof is still intact and the area is not filled with rubble. The bags are placed in the mine through the borehole and are then filled with grout. As the cylindrical bag fills it forms a column that provides a point support (Figure 4).

A test of this method was done at a site in Ohio. The site was at a slope entry to an abandoned mine in the Pittsburgh Coal Seam. The mine void was about 12 meters below a road. The bags were inserted through 20 cm diameter holes.

Ten synthetic bags were installed that were 3.6 meters high and 1.8 meters in diameter. Each bag was folded to form a narrow elongated bundle and placed in a split 15 cm diameter PVC pipe for passage through the borehole.

The bags had internal straps to control bulging. To further minimize stress on the bags a low density grout was used. The grout mix was a cellular concrete mixed on-site by a mobile batch plant. The mix consisted of cement fly ash, a foaming agent and water. The grout was pumped into the bags through a 7.6 cm diameter injection pipe. The lower section of the injection pipe was designed to remain in

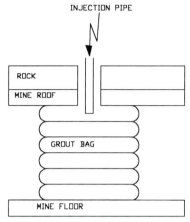

Figure 4 : Grout Bag

the borehole to hold the neck of the bag in place after filling.

The bags were filled in one meter stages and before completing the first stage the
bag was lifted up to allow the bottom of the bag to spread smoothly on the mine
floor. During the final stage the injection was slowed to allow the grout to form as
broad a roof contact as possible.

At this site the bags were able to provide roof support when placement was done
in a controlled manner. This method significantly reduces the volume of grout
required to provide mine roof support.

Sodium Silicate Grout

The placement of grouts to form supporting columns underwater frequently is
inadequate because the grout quickly becomes diluted from the movement of the
surrounding water. Building a column that will provide roof contact using a
minimum amount of grout requires a low-slump, high-viscosity mixture. When a
grout of this type comes in contact with water, the mix starts to become diluted as
water passes into the grout matrix. Dilution reduces the viscosity and self
supporting properties of the grout, preventing the formation of a column and
resulting in a pile exhibiting a low angle of repose. This also lowers the strength in
the grout and it is unable to provide support to the mine roof. To compensate for

this problem a technique was tested to pass the grout stream through a curtain of sodium silicate during placement to form a calcium silicate gel barrier on the grout surface that limits the dilution of the grout.

A field scale study was done to test the underwater formation of grout columns (Reifsnyder, Brennan, and Peters 1988). The study used swimming pools to simulate underwater mine void conditions. The grout used was six parts fly ash (Mercer Class F) and one part cement (Portland type two) by volume. These materials were combined in a grout mixer and tap water was added to bring the total solids of the grout to 72% by weight. The silicate used was a 3.22 : 1 weight ratio (SiO_2 to Na_2O) sodium silicate liquid having a solids content of 37.6% by weight. During tests it was determined that the injection pipe should be located near the point of placement (tremied) to minimize fall of the grout through the water column to minimize dilution (Figure 5).

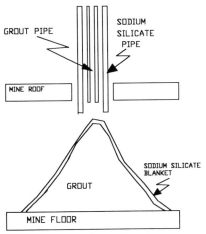

Figure 5 : Sodium Silicate
Injection Pipe

The study found that the control column without a sodium silicate cover attained a height of only 46 cm and had an average angle of repose of 5 degrees. In the other tests, with silicate being introduced at the nozzle, the grout was able to form columns having a high angle of repose, maximizing structure height while minimizing raw material usage. The average angle of repose of the silicate-enhanced columns was at least 5.5 times greater than the control. The inclusion of silicate into the grouting operation also was found to enhance the stability of fresh grout placed underwater, preventing separation of the grout components while improving the stiffness and self-supporting qualities. The silicate was also found to improve the rate of grout setting and the final compressive strength of the column.

Cylindrical Grout Columns

A method of providing point support in abandoned coal mines was evaluated using boreholes to build a cylindrical column with low slump grout from the mine floor to the roof (Burnett, El-Korchi, and Burnett 1993).

A 20 cm diameter borehole was used to access the mine void. A grout pipe with a flexible trunk was lowered to the mine from a frame that could raise and lower the pipe in the borehole. There was also a motor driven rotating mechanism to rotate the pipe and a flexible trunk on the down-hole end of the pipe that was bent with an 46 cm radius while grout was being placed (Figure 6). The cylindrical column of grout was 2.4 meters in diameter at its contact point with the mine roof. This type of hollow grout column provides significant support to the mine roof over a large area and minimizes the volume of material required.

SUMMARY

Abandoned underground coal mines are located throughout the coal fields in the United States. Structures and facilities above these mines need to be protected from subsidence caused by the collapse of the mine workings. To protect the public from subsidence and other adverse effects of coal mining an Abandoned Mine Lands Reclamation Fund was established under Public Law 95-87. This fund is used to finance subsidence control projects that backfill the mine voids to control subsidence. Grout is frequently used to backfill and stabilize these areas. However, depending on the conditions in the mine the placement of the grout may require unique equipment or customized grout materials. Abandoned mines are generally inaccessible, the areas requiring stabilization are beneath surface structures, the mines may be flooded, filled with rubble or the mine floor may slope. In addition to special considerations in grout placement the areas requiring roof support may be extensive and filling the voids in the entire mine are cost prohibitive. Special grout placement techniques are available including controlled grout columns, synthetic grout bags, and chemical additives to control grout setup.

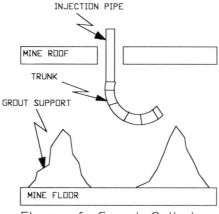

Figure 6: Grout Cylinder

REFERENCES

Ackenheil, A.C., and Dougherty, M.T., (1970), "Recent Developments in Grouting for Deep Mines." *Journal of Soil Mechanics and Foundations Division Proceedings, ASCE, Volume 96, No. 1*, 251-261.

Burnett, M., El-Korchi, T., and Burnett, J., (1993), "Construction of Large Diameter Concrete Columns Through Boreholes", prepared by Burnett Associates for the Bureau of Mines, United States Department of the Interior.

Craft, J.L., (1990), "Classification of Mine-Related Subsidence East of the Mississippi River, U.S.A." *Association of Engineering Geologists Annual Meeting Proceedings*.

Gray, R.E., and Meyers, J.F., (1970), "Mine Subsidence and Support Methods in the Pittsburgh Area." *Journal of Soil Mechanics and Foundations Division*

Proceedings, ASCE, Volume 96, No. 4, 1267-1287.

Michael, P.R., Lees, A.S., Crandall, T.M., and Craft, J.L., (1987), "Controlled Grout Columns : A Point-Support Technique for Subsidence Abatement." *Association of Engineering Geologists Symposium Series, Number 4*, 111-125.

Office of Surface Mining, (1982), "Abandoned Mine Lands Reclamation Control Technology Handbook." *U.S. Government Printing Office.*

Office of Surface Mining, (1998), "1997 Annual Report." *U.S. Government Printing Office.*

Office of Technology Assessment, (1979), "The Direct Use of Coal." *U.S. Government Printing Office.*

Reifsnyder, R.H., Brennan, R.A., and Peters, J.F., (1988). "Sodium Silicate Grout Technology for Effective Stabilization of Abandoned Flooded Mines." *Proceedings Mine Drainage and Surface Mine Reclamation Conference,* Pittsburgh, PA, Volume II, 390-398.

Stabilization and Closure Design of a Salt Mine

Nasim Uddin, Ph.D., P.E., M. ASCE [1]

Abstract

This paper describes the design for the stabilization and the closure of the Detroit Salt Mine which includes design of the stabilization measures in the mine to ensure long term mine stability, and concrete plugs in two mine shafts to prevent long-term leakage into the mine and subsequent solutioning of the salt. The design includes grouting in the mine at two floor levels, grouting in and around the shaft plugs including two-stage cement and chemical curtain grouting in the rock surrounding the plug and high pressure contact grouting at the concrete-soil and concrete-rock interface. The paper addresses the selection of plug locations, plug geometry, concrete mix design, cooling of the concrete and grouting of the rock and rock/concrete interface. Lack of proper maintenance and repair during the life of the 90-year old facility and an aggressive subsurface environment, presented many unusual design issues. The paper describes how the mine stabilization and closure design addressed the issues and ensured that ground water was controlled at all phases of the construction.

Introduction

Salt production at the Detroit Mine stopped in 1983. Since then various attempts have been made to find a use for the mine, including storage of waste products, natural gas and compressed air for power generation. However, these uses for the mine have not proved economically feasible, and in order to minimize future maintenance of the mine, the owner hired Acres International to design a closure system for the mine. The room and pillar mine workings are stable with no signs of salt movement or rock distress. However, significant leakage occurs at the shafts. The main objective for the closure design is to seal the shafts and prevent water inflow into the mine. Long term seepage into the closed mine could result in solutioning of the salt

----

[1] Assistant Professor, University of Evansville, 1800 Lincoln Avenue, IN 47722

and progressive collapse, potentially to the surface.

The intent of the design of the closure system is to seal the shafts to prevent leakage into the mine and preserve the current mine stability. Closure of the mine will eliminate the long term maintenance and repairs needed to provide safe access for inspection and pumping. Detailed design is currently completed and bidding for the project will start soon.

Mine Development

The Detroit Mine is located in the Detroit, Allen Park and Melvindale Townships south of the River Rouge in Detroit. The mine is adjacent to a large Ford Motor Co. Plant and extends under a mixture of residential and heavy industrial properties. A major railroad crosses above the mine in a northeast direction. The mine construction was started in 1906 with the sinking of Shaft No. 1 which took five years to complete. Very difficult ground and artesian water conditions were encountered as described in a paper by Albert H. Fay (Fay 1911). Construction of a second shaft followed, believed to have been completed in 1922. Over the next 8 to 10 years, extensive repairs were performed on both shafts. Two 42-inch diameter pipes were installed in shaft No. 1 and the shaft was backfilled around the pipes with concrete. It is believed that this was an attempt to control leakage into the shaft.

Extensive repairs were made to the No. 2 shaft. Sections of the concrete lining were replaced and at some time later, a brick lining was placed in the shaft. The No. 1 shaft was used for backup and emergency personnel access with a two-level man cage in each of the 42-inch diameter inner shaft pipes. No. 2 shaft was used for hauling salt. The bottom of the No. 2 shaft was equipped with loading pockets. After some initial mining at the A salt level, salt production was switched to the B level and continued at that level for the remainder of the mine operation. Salt was mined from about 1910 to 1983 by a room and pillar method with about 50 percent extraction. The layout of the mined areas is irregular, to suit mining rights and sensitive property, such as a railroad corridor. The mine covers approximately two square miles. The mine is dry with no water leakage except at the shafts. Some pools of water occur within the mine. This water is thought to be associated with condensation from the ventilation system or connate water from the salt itself.

Site Geology

The geology of the immediate area of the Detroit Salt Mine was derived from logs of the mine shafts, exploratory core holes and nearby oil, gas and brine wells. Figure 1 shows the generalized geological section of the strata underlying and penetrated by the excavation of the No. 1 and No. 2 shafts. At the mine site, surficial geology consists of 86 ft of unconsolidated glacial deposits (mostly blue clay with scattered cobbles and boulders). Underlying the glacial deposits is approximately 330 ft of Devonian limestone and dolomite. Next below is 113 ft of water bearing Sylvania sandstone, below which lies another 344 ft of dolomite that includes a few shaly beds toward the base. Next below is approximately 1,000 ft thickness of interbedded dolomite, salt, shale and anhydrites characteristic of the Salina Group evaporites. The salt units at the mine generally strike to the ENE and dip or slope gently to the north

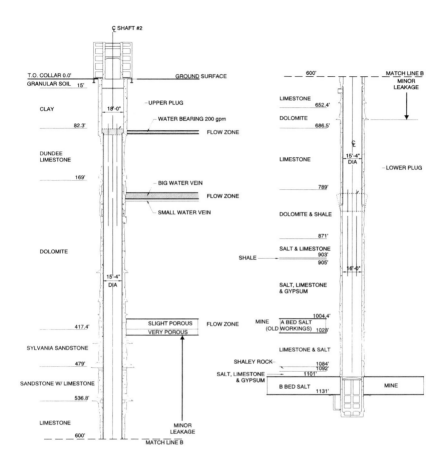

SHAFT NO. 2
DETROIT SALT MINE

Figure 1. Generalized geological section

at approximately 1.4 ft per 100 ft.

Ground Water

The groundwater hydrology in the vicinity of the Detroit Salt Mine was determined from logs of the mine shafts and most recently in the drilling of DH-1, a salt disposal well installed in 1994. From the surface down to a depth of 680 ft, several major water bearing zones were encountered. Artesian conditions are encountered at depths of 85 to 100 ft (top of rock) and 170 to 190 ft (Lucas Dolomites). The groundwater is generally of poor quality and highly charged with H_2S at depths below 80 ft. Seepage of groundwater into the shaft from the shaft penetrations is currently collected by a system of gravity feed lines into sumps and tanks at mine level and subsequently pumped to the ground surface for disposal.

Currently leakage rates are 30 gpm for shaft No. 1 and 20 gpm for shaft No. 2. Periodic measurements indicate that these rates are increasing. Historical data indicates that the aquifers penetrated by the shafts are under artesian head of about 20 ft as measured at the ground surface. This is consistent with recent report (Raven 1992) of "super normal" pressures up to 1.7 times hydrostatic pressure measured in similar rock strata in Southern Ontario. Artesian flows were encountered in borehole DH-1 where flows to the surface of up to 1000 gpm were estimated at a hole depth of 170 ft.

Mine Closure

The primary objective in the design of the mine closure was to stabilize the salt pillars at two floor levels, and seal the shafts and prevent leakage into the mine which would result in solutioning of the salt and result in mine instability. To meet this primary objective, it was decided to construct concrete plugs in each shaft. The plugs were located as high as possible in the shaft to minimize the hydrostatic head on the plugs at the same time ensuring that the plugs were below any potential leakage zones. The plug location selected was within a competent dolomite and shale stratum 300 ft above mine level. Because of the possibility of artesian flow from the shaft once the lower plugs are installed, a second plug is required at the top of the shaft. The upper plugs also serve as protection caps to the top of the shafts.

Factors Considered in the Design of Plugs

One of the most important factors considered in deciding where to place a plug is the condition of the surrounding rock. Detailed geologic evaluations are carried out to locate the plug in the rock (a) which is free from geological disturbances which could provide a leakage path for water; (b) not in or near the fracture zones of highly stressed ground resulting from mining excavations; and (c) not in the ground likely to be affected by subsequent ground movement resulting from external disturbances.

Figure 2 shows the arrangement of the lower plugs. This concrete plug is unreinforced and incorporates a taper to ensure compressive stress over the entire rock/concrete contact area. The resistance to the applied hydrostatic pressure is achieved through mechanical interlock with the rough excavation face of the surrounding rock. Garrett and Campbell Pitt (1961) considered plug length to be governed more by leakage resistance around the sides and through the surrounding

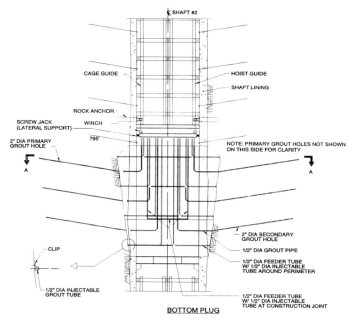

BOTTOM PLUG

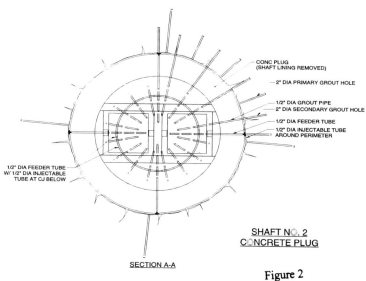

SECTION A-A

SHAFT NO. 2
CONCRETE PLUG

Figure 2

rock than by structural strength. The long plug length required for leakage resistance
also ensures low shear and bearing stresses at the concrete to rock interface. The plug
design required that the concrete lining be completely removed to ensure that leakage
paths at the rock lining interface were intercepted. The rock excavation for the tapered
plug also ensured that rock disturbed by the blasting for the original shaft excavation
would be removed.

The following factors are considered when evaluating the stresses in, and
strength of, the plugs: (a) concrete compressive strength; (b) the early stage
development of strength; (c) the shear or bearing stress at the plug to rock interface;
(d) the pore water pressure in the concrete; and (e) the possible end spalling of the plug
due to high stresses set up by ground pressure. Early stage development strength is
very important because it is essential that plugs develop their specified strength
without any detrimental effects occurring from shrinkage, thermal changes or ground
pressure. Steps are, therefore, taken to overcome these factors, to protect the integrity
of the concrete and to minimize the leakage path through the plugs. Since the length
of the plug is calculated based on the minimum leakage path, it is observed that a
longer length of plug is provided than is necessary for structural strength purposes.
Note that knowledge of pore-water pressure behavior within a concrete plug is very
limited. Therefore, a pressure gradient is assumed to exist from the hydrostatic
pressure at the face in contact with the impounded water to zero at the other end.
Garrett and Campbell (1959) proposed an approach which is based on the observation
that the resistance of the plug to passage of water either along its contact with rock or
through the adjacent fractured rock depends on the length of the plug and the resistance
of the rock to the passage of water. They observed that these two factors are
interrelated by the pressure gradient through the rock as the linking medium. They
also determined the minimum length of the plug if the contact between plug and rock
was ungrouted, as follows:

$$l = \frac{Pressure \ gradient \ among \ two \ faces}{20.8 \ lb/inch^2/ft}$$

It was determined that a 20 ft long plug was necessary to satisfy the above criteria.
However, due to the uncertainties associated with the shaft and rock conditions and the
importance of long term performance of the plugs, an additional factor of safety of 1.5
was applied to the plug lengths. Provisions were also made for grouting of the
surrounding rock and the rock/concrete interface as additional safety measures.

Control of Water During Construction

A crucial element of the design of the shaft plugs was control of the seepage
into the mine and shafts during the construction of the plugs. The design incorporated
a drainage collection system and temporary pumps to keep the construction area dry
until the plugs could be concreted and grouted. Back up systems were specified to
ensure safety of workers in the shafts. Design provisions were also made for the
dewatering of the shafts for the upper plugs, should the shafts flood and fill before
construction of the upper plugs is completed.

Concrete Placements

Considerations are made to provide the right balance of ingredients to suit the particular mode of transportation and placing being used and to make sure that the optimum design is achieved. In addition to strength, the most important factor for the mix design is to obtain the correct workability. For this underground work, with its restricted placing environment, it is essential that high workability mixes are used. A mix design is, therefore, developed by the inclusion of: (a) plasticizing admixtures for transporting and placing; and (b) cement replacement materials to minimize the thermal effects which are discussed below. The large volume of concrete required for mass filling (310 cu.yd. for shaft #1 and 422 cu. yd. for shaft #2) can be subjected to detrimental thermal effects during setting. The resulting shrinkage is dependent upon the amount of cement included in the mix. Internal build up of heat due to the cement hydration process induces a high thermal stress. On cooling, thermal cracking may result and the integrity of the structure would be impaired. The three elements of the effective temperature control program are used for the project: 1.) cement content control, where particular type and amount of cements material (Type IV cement with fly ash and 2 inch course aggregate) is used to lessen the heat generating potential of the concrete; 2.) pre-cooling, where cooling of ingredients (set to 50°F by removing approximately 25,000 btu from the concrete components using liquid nitrogen) achieves a lower concrete temperature as placed in the plug; 3.) post-cooling, where removing heat from the concrete with embedded cooling coils limit the temperature rise in the plug. Figure 3 shows the post cooling measures for the plug concrete to limit the concrete temperature rise as minimal as possible.

Grouting

The intent of the grouting program was to stabilize the both mine floor levels A and B, reduce the permeability of the rock surrounding the plug and to fill all voids at the contact between the rock and plug concrete due to inadequate concrete placement or shrinkage. A comprehensive grouting program was specified as follows:

A. Cement grouting: Work included :

(a)   Consolidation and backfill grouting of voids and unstable zones below pillars and concrete piers at A level and B level in the mine

(b)   Primary curtain grouting at the concrete plugs in both shafts nos. 1 and 2

(c)   Sealing and grouting of holes, penetrations, dewatering pipes, coolant pipes and grout tubes at and in the shaft plugs.

Figs. 4 and 5 show extent of new voids developed in mine floor level A and B respectively (since the repair in 1980) and grout pipe installed (in void) to inject cement grout to fill the voids. Also included exploratory holes to be used as grout holes if required. Also shown some detail of grout hole locations, and approximate area of salt pillars requiring restoration by means of grout placement. Figs.6a, 6b and 6c show some typical sections at B salt level with grout holes to extend approximately

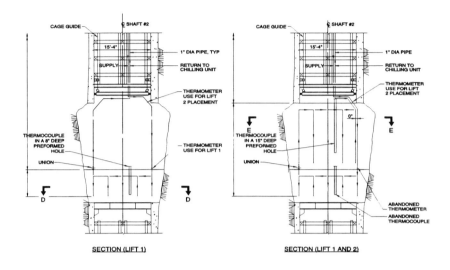

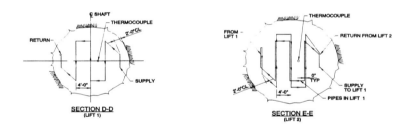

CONCRETE PLUG
COOLING SYSTEM

Figure 3

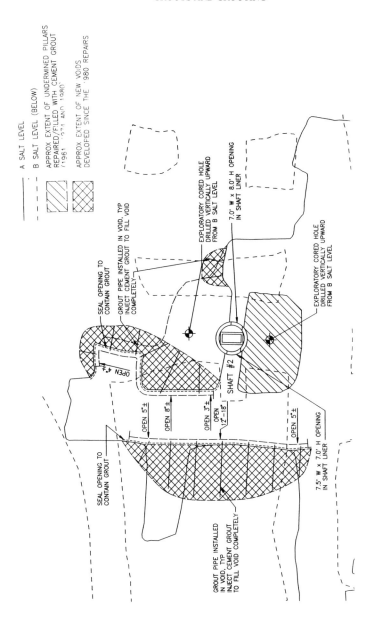

Figure 4. Plan showing extent of mine stabilization measures at 'A' floor level

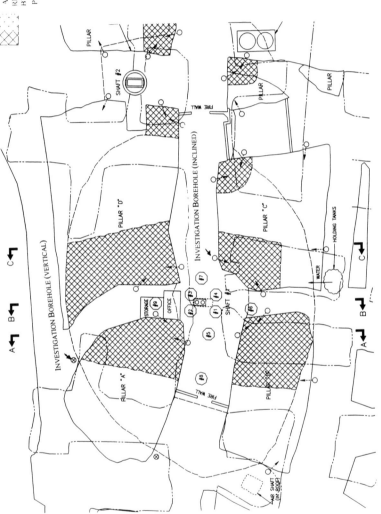

Figure 5. Plan showing extent of mine stabilization measures at 'B' floor level

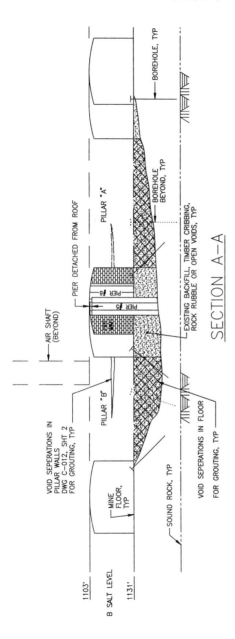

Figure 6(a). Section A-A showing approximate area of salt pillars requiring restoration by means of grout placements at 'B' floor level

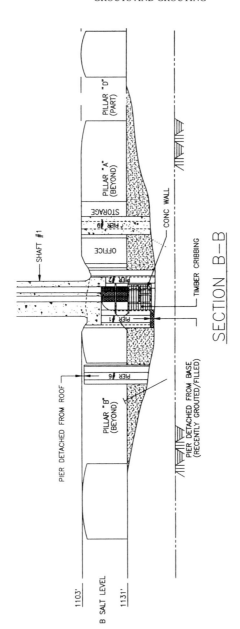

Figure 6(b). Section B-B showing approximate area of salt pillars requiring restoration by means of grout placements at 'B' floor level

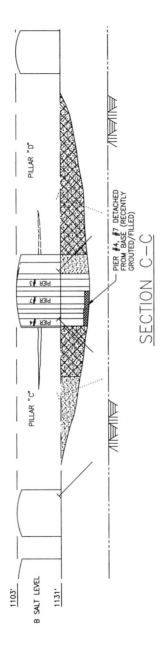

Figure 6(c).   Section C-C showing approximate area of salt pillars requiring restoration by means of grout placements at 'B' floor level

B. Chemical Grouting: Work included:
(a)   Contact grouting of the rock/concrete interface after construction of the plug and cooling to ambient temperature
(b)   Second stage rock grouting program using an acrylamide grout when the contact grouting is completed.
Figure 5 shows some details for the contact and secondary curtain grouting program for the shaft no. 2.

Grout Materials -
        Contact Grouting: Polyurethane grout will be used for contact grouting of shaft plugs which will involve injection of grout to fill annular space at rock/concrete interface or at concrete construction joint. If water is present at the rock/concrete interface, a water reactive, flexible, hydrophilic, MDI-based polyurethane prepolymer such as Polyurethane resin will be used. If no water is present at the rock/concrete interface, a two component polyurethane elastomer with hydrophillic antennas such as Polycast EXP as distributed by Multiurethanes, Inc. will be used.
        Secondary Curtain Grouting: Secondary curtain grouting will be performed in predrilled holes surrounding the plug to fill open fractures or voids in the rock structure in order to decrease the permeability of the rock mass. The intent of the secondary curtain grouting is to reduce the rock mass permeability to 1×10^{-7} cm/sec. Following grout ingredients will be used: 1. Mono - acrylamide powder containing the cross linker, 2. Activator - tri - ethanol-amine (T+), 3. Initiator - ammonium persulfate (AP), 4. Inhibitor - potassium ferricyanide (KfeCN), 5. Buffer - sodium bicarbonate, 6. Water, and 7. Dye. Acrylamide powder containing 95% acrylamide monomers and 5% cross linker will be mixed with water to a 40% solution offsite and will be delivered to site in solution form. Grout will be mixed on site at the point of injection by combining two components and will be pumped at the hole in equal volumes resulting a 20% acrylamide solution.

Execution: A porous tube ½-inch diameter with spiral steel reinforcement (to prevent collapse during placement of concrete) and woven membrane which will not allow passage of cement particles but will allow easy passage of polyurethane grout will be used as contact grout injectable tubes. ½-inch internal diameter steel tubes rated for 1,000 psi internal pressure will be used for feeder tube for polyurethane grout. Injectable grout tube will be secured directly to rock or concrete surface with fastening clips to maintain tube in contact with rock or concrete during the placing of concrete. Sufficient tension tube will be used as to not allow movement of the tube during concrete placement while maintaining the tube in contact with the concrete or rock surface. Each injectable tube will form a continuous ring or loop around the plug.
        ½-inch diameter stainless steel pipe rated for two times maximum pump pressure will be used for grout pipes for acrylamide grout. Equipment will include two mixing reservoirs along with separate grout lines for each component of acrylamide, flow meter, transfer pumps, two-component piston pumps, X-Y recorder etc. Grout pipes will be installed from secondary curtain grouting holes to 2 ft above plug

45 ft. below the mine floor. Fig. 2 shows primary curtain grouting at the concrete plugs in shafts. These holes are 2 inch in diameter and 20 ft. deep inside the rock making approximately 10 degrees with the horizontal.

Grout Materials: Grout mixes to be used for void filing at mine level are described in Appendix A. Note that Portland Cement Type I and III as specified in ASTM C150 and Brine water (95% to 100% salt saturated) along with admixtures will be used for grout applications. Brine will be used in order to prevent solutioning of salt and will be manufactured on site using fresh, potable water combined with "fine C" salt available at mine level B and will be filtered to remove particulate matter greater than No. 200 sieve size. Some applications for filling large voids in the mine may be amenable to pre-installing a suitable aggregate (ASTM C33 Size No. 8 or Michigan DOT 8.02.03 Series 28/B) for pre-placed aggregate grouting.

Execution: The equipments for grouting will include High-shear mixers (1,400 to 2,000 rpm), a reservoir containing a low-adjustable-speed (40 to 200 rpm), paddle mixers (to prevent flocculation and segregation of the mix), Helicoidal screw pumps (to deliver grout at a minimum pressure of 150 psi which could be needed to compensate for the line losses when utilizing grout with high cohesion/viscosity), a double-line circulating system, multiple- hole grouting header, gages, flow meters, X-Y recorder, packers or seals and measuring devices. Percussion type drilling equipment will be used for drilling holes. If drilling is difficult because of position or rock condition, rotary drill with a diamond bit will be used. Stage grouting (both up- and down- hole) techniques will be used. Maximum length of grout stage will be 10 ft. There are various approaches to evaluate the suitability and effectiveness of cement-based suspension grouts injected in a particular formation in soil or rock grouting. The amenability theory, now commonly used in North America, will be used as the basis of assessing the suitability and effectiveness of the suspension grouting specified herein. Grout will be mixed in batches of suitable size to permit grout formulations to be varied quickly to suit the response of the formation and achieve maximum amenability. Since some applications of this grout are intended to be restricted or confined to a predetermined lateral spread, depending on the amenability of the formation and the amount of grout already injected into that specific zone, a viscosity modifier will be used, only marginally influencing the grout-rheology at high-shear rates, but causing a substantial increase in viscosity in the cement grout when moving slowly or at rest. Furthermore, a retarder, such as Polycast LSR, or approved equivalent, will be used to allow the formation to be reaccessed after 24 hours.

All grout holes/zones will be grouted to refusal. A zone or hole is brought to refusal once the grout flow rate to the hole/zone has been registered to be less than 0.10 gallons per minute (gpm) for a minimum of 20 minutes at the maximum allowable effective grouting pressure. Maximum allowable grouting pressure for a hole/zone will be one psi per foot of depth or surface cover at the interval being grouted, or will be determined as a result of trials. If a back pressure exists after the grouting of a hole is completed, the hole will be capped until the pressure falls to a negligible amount (because of the concern for safety due to over pressurizing).

concrete surface. Grout pipes will be inserted 2 ft into the grout hole and the annulus between borehole and grout pipe will be sealed with fast set cement mortar to prevent concrete from entering borehole. Seal will be adequate to prevent the permeation of the contact-grout into the secondary grout holes. Surface of the mortar will be treated with a polyurethane or epoxy coating to prevent permeation of the contact grout. In-situ hydraulic conductivity tests will be performed with real time monitoring. Gel time will be determined from the tests. Holes with similar in-situ permeability coefficient will be identified for multiple hole grouting program. Appropriate precautions will be taken with respect to acrylamide toxicity. All individuals entering the work area where the acrylamide grouting is performed will wear appropriate clothing, respirators and safety gears. Any contact of acrylamide with the skin must be promptly and throughly flushed with water.

Acknowledgments

A number of Acres International Corporation personnel participated in this project including the author and would like to acknowledge their contributions: Project manager Nigel J. Bond and Staff Geologist Gary W. Page.

References

Fay, A.H., "Shaft of Detroit Salt Company," The Engineering and Mining Journal, p.g. 565-569, March 18, 1911.

Garrett, W.S. and Campbell Pitt, L.T., "Design and construction of undergorund bulkheads and water barriers", Seventh Commonwealth Mining and Metallurgical Congress, Johannesburg, 1961.

Raven, K.G., et. al.,"Supernormal Fluid Pressures in Sedimentary Rocks and Southern Ontario-Western New York State," Canadian Geotechnical Journal, Vol. 29, pg. 80-93, 1992.

Appendix A

SUGGESTED GROUT MIX FORMULATIONS FOR BACKFILL GROUTING AT MINE LEVELS 'A' AND 'B'

1. Thin Mix (Low Viscosity Mix Formulation)

    100   Part brine (saturated in NaCl)

      4   Parts bentonite (or in the form of 40 parts from a prehydrated bentonite slurry at a 10% concentration which has been prepared in a high-shear mixer in advance of grouting. The amount of water in the slurry needs to be deducted from the amount of brine.)

0.25   Parts naphthalene sulphonate (if <u>no</u> retardation is required)

   1   Part Polycast LSR (if retardation is required)

 100   Part Type 1 cement

  35   Part Class F flyash

 0.1   Part viscosity modifier: Botor X10 (when required)

2.   <u>Medium Mix (Medium Viscosity Mix Formulation)</u>

 100   Parts brine

   4   Parts bentonite

 0.5   Parts naphthalene sulphonate (if <u>no</u> retardation is required)

1.25   Parts Polycast LSR (if retardation is required)

 100   Parts Type 1 cement

67.5   Parts Class F flyash

 0.1   Part viscosity modifier: Botor X10 (when required)

3.   <u>Thick Mix (High Viscosity Mix Formulation)</u>

Medium mix formulation to be thickened by gradually decreasing brine content and/or adding viscosity modifier in small increments.

Note: All mix proportions are provided in terms of weight.

High Flow Reduction in Major Structures:
Materials, Principles, and Case Histories

D.A. Bruce[1], M.ASCE, A. Naudts[2,] and W.G. Smoak[3]

Abstract

The paper firstly presents a structured classification of the four different categories of grouting materials: particulate, colloidal, solutions, and miscellaneous. Three case histories are then summarized to illustrate the application of these materials to produce optimum results in projects where the goal is to reduce or eliminate high seepage flows, usually at high hydrostatic pressures. The case histories are drawn from recent works conducted by the authors at Dworshak Dam, ID, Tims Ford Dam, TN, and at a potash mine in New Brunswick, Canada. Conclusions are drawn on the eight elements common to the achievement of a satisfactory result in such programs.

Introduction

One of the most difficult challenges facing the grouting industry is the reduction or elimination of high volume water inflows into or through major civil engineering structures such as dams, tunnels, and quarries. Often these flows are occurring at high velocities, under high heads, and in locations which render treatment logistically and practically very awkward. Of particular concern are those situations such as karstic limestone formations, where there are often networks of orifices, as well as zones with potentially erodible or soluble material that must also be treated to prevent future reoccurrence of the problem.

[1]Principal, ECO Geosystems, Inc., P.O. Box 237, Venetia (Pittsburgh) PA 15367 phone: (724) 942-0570 fax: (724) 942-1911; email: eco111@aol.com
[2]General Manager, ECO Grouting Specialists, Ltd., Cheltenham (Toronto), ON.
[3]Principal, ECO Structural Systems, LLC, Lakewood (Denver) CO.

Over the years, many existing major structures have been treated by remedial grouting operations, but with varying degrees of success. One of the major reasons contributing to this erratic performance has been the inappropriate selection of the grouting materials. The first part of this paper provides a generic classification of the major families of grout that can be used in such applications.

The second part of the paper provides summary accounts of three major recent case histories, which illustrate a systematic approach that can be adopted towards analyzing the issues and problems, executing the work, and optimizing and verifying the results.

Generic Classification of Grouting Materials

There is a plethora of grouting materials available which, given the broad range of their chemical compositions, and trade names, can be bewildering even to specialists in the field. In addition, it should also be noted that different placement methods and techniques will be required for different materials: the "conventional" staging processes used in the construction of cementitious grout seepage barriers may not be suitable, for example, to the special intricacies of injecting hot melts (e.g., bitumen) and their associated materials. More details on the materials introduced below can be in found in Naudts (1996) and Bruce et al. (1997). A companion paper by Bruce (1992) describes the drilling and grouting construction principles used in dam rehabilitation, although these can be extended to cover other applications.

Basis of Classification

There are four categories of materials, listed in order of increasing rheological performance and cost:

1. Particulate (suspension or cementitious) grouts, having a Binghamian performance.
2. Colloidal solutions, which are evolutive Newtonian fluids in which viscosity increases with time.
3. Pure solutions, being nonevolutive Newtonian solutions in which viscosity is essentially constant until setting, within an adjustable period.
4. "Miscellaneous" materials.

Category 1 comprises mixtures of water and one or several particulate solids such as cement, pozzolans, clays, sand or viscosity modifiers. Such mixes, depending on their composition, may prove to be stable (i.e., having minimal bleeding) or unstable, when left at rest. Stable, thixotropic grouts have both cohesion and plastic viscosity increasing with time at a rate that may be considerably accelerated under pressure.

Category 2 and 3 grouts are now commonly referred to as solution or chemical grouts and are typically subdivided on the basis of their component chemistries, for example, silicate based (Category 2), or resins (Category 3). The outstanding rheological properties of certain Category 3 grouts, together with their low viscosities, permit permeation of soils as fine as silty sands ($k = 10^{-4}$ cm/s).

Category 4 comprises a wide range of relatively exotic grout materials, which have been used relatively infrequently, and only in certain industries and markets. Nevertheless, their importance and significance is growing due to the high performance standards which can be achieved when they are correctly used. The current renaissance in the use of hot bitumen grouts is a good example.

Category 1: Particulate Grouts

Due to their basic characteristics, and relative economy, these grouts remain the most commonly used for both routine waterproofing and ground strengthening. The water to solids ratio is the prime determinant of their properties and basic characteristics such as stability, fluidity, rheology, strength, and durability. Five broad subcategories can be identified:

1. Neat cement grouts.
2. Clay/bentonite-cement grouts.
3. Grouts with fillers.
4. Grouts for special applications.
5. Grouts with enhanced penetrability.

It should be borne in mind that many particulate grouts are unsuited for sealing high flow, high head conditions: they will be diluted or washed away prior to setting in the desired location. However, the recent developments in rheology and hydration control technologies, and the advances made in antiwashout additives have offered new opportunities to exploit the many economic, logistical, and long term performance benefits of cementitious compounds (Gause and Bruce, 1997). Low mobility grouts ("compaction grouts") can be classified in the third subgroup, and can be very beneficial in flow reduction under appropriate conditions as noted below.

Category 2: Colloidal Solutions

These comprise mixtures of sodium silicate and reagent solutions, which change in viscosity over time to produce a gel. Sodium silicate is an alkaline, colloidal aqueous solution. It is characterized by the molecular ratio R_p, and its specific density, expressed in degrees Baumé (Bé). Typically R_p is in the range 3 to 4, while specific density varies from 30 to 42 Bé. Reagents may be organic or inorganic (mineral). The former cause a saponification hydraulic reaction that frees acids, and can produce either soft or hard gels depending on silicate and reagent concentrations. Common types include monoesters, diesters, triesters, and aldehydes, while organic acids (e.g., citric) and esters are now much less common. Inorganic reagents contain cations capable of neutralizing silicate alkalinity. In order to obtain a satisfactory hardening time, the silicate must be strongly diluted, and so these gels are typically weak and therefore of use only for waterproofing. Typical inorganic reagents are sodium bicarbonate and sodium aluminate.

The relative proportions of silicate and reagent will determine by their own chemistry and concentration the desired short- and long-term properties such as gel setting time, viscosity, strength, syneresis, and durability, as well as cost and environmental acceptability.

In general, sodium silicate grouts are unsuitable for providing permanent barriers against high flow/high head conditions, because of their relatively long setting time (20 to 60 minutes), low strength (less than 1 MPa) and poor durability. This is a different case from using sodium silicate solution (without reagent) to accelerate the stiffening of cementitious grouts - a traditional defense against fast flows.

Category 3: Pure Solutions

Resins are solutions of organic products in water, or a nonaqueous solvent, capable of causing the formation of a gel or foam with specific mechanical properties under normal temperature conditions and in a closed environment. They exist in different forms characterized by their mode of reaction or hardening:

- Polymerization: activated by the addition of a catalyzing element (e.g., poly-acrylamide resins, water reactive polyurethanes).
- Polymerization and Polycondensation: arising from the combination of two components reacting in stoechiometric proportions (e.g., epoxies, aminoplasts, two component polyurethanes, vinyl esters).

Mostly, setting time is controlled by varying the proportions of reagents or components. Resins are used when Category 1 or 2 grouts prove inadequate, for example when the following grout properties are needed:

- particularly low viscosity.
- very fast gain of strength (a few hours).
- variable setting time (few seconds to several hours).
- superior chemical resistance.
- special rheological properties (pseudoplastic).

- resistance to high groundwater flows.

Resins are used for both strengthening and waterproofing in cases where durability is essential, and the above characteristics must be provided. Four categories can be recognized: acrylic, phenolic, aminoplastic, and polyurethane (Table 1). Chrome lignosulfonates are not discussed, being, according to Naudts (1996), "a reminder of the dark, pioneering days of solution grouting" on account of the environmental damage caused by the highly toxic and dermatitic components.

Type of Resin	Nature of Ground	Use/Application
Acrylic	Granular, very fine soils	Waterproofing by mass treatment
	Finely fissured rock	Gas tightening (mines, storage) Strengthening up to 1.5 MPa Strengthening of a granular medium subjected to vibrations
Phenol	Granular, very fine soils	Strengthening
Aminoplast	Schists and coals	Strengthening (by adherence to materials of organic origin)
Polyurethane	Large voids	Formation of a foam that forms a barrier against running water (using water-reactive resins) Stabilization or localized filling (using two-component resins)

Table 1. Uses and applications of Resins (AFTES, 1991).

Of these four subclasses, only the two groups of polyurethanes are usually appropriate for remedial grouting given cost, performance and environmental implications:

- Water-reactive polyurethanes: Liquid resin, often "reactively diluted" or in a plasticizing agent, typically with added accelerator, reacts with groundwater to provide either a flexible (elastomeric) or rigid foam. Viscosities range from 50 to 1,000 cP (at $25\,^{\circ}$C). There are two subdivisions:

1)   Hydrophobic - react with water but repel it
        after the final (cured) product has been
        formed.
2)   Hydrophillic - react with water but continue to
        physically absorb it after the chemical reaction
        has been completed.

•   Two component polyurethanes: Two compounds (polyol and
        isocyanate) in liquid form react to provide either a rigid foam
        or an elastic gel. Such resins have viscosities from 100 to
        1,000 cP and strengths as high as 2 MPa. A thorough
        description of these grouts was provided by Naudts (1996).

Category 4: Miscellaneous Grouts

The following grouts are essentially composed of organic compounds or resins. In
addition to waterproofing and strengthening, they also provide very specific qualities
such as resistance to erosion or corrosion, and flexibility. Their use may be limited
by specific concerns such as toxicity, injection and handling difficulties, and cost.
Categories include hot melts, latex, polyesters, epoxies, furanic resins, silicones, and
silacsols. Some of these (e.g., polyesters and epoxies) have little or no application
for ground treatment. Others such as latex and furanic resins are even more obscure
and are not described.

For certain cases in seepage cut off, hot melts can be a particularly viable option.
Bitumens are composed of hydrocarbons of very high molecular weights, usually
obtained from the residues of petroleum distillation. Bitumen may be viscous to hard
at room temperature, and have relatively low viscosity (15 to 100 cP) when hot (say
over 200°C). They are used in particularly challenging water-stopping applications
(Bruce, 1990a and b; Naudts, 1996), remain stable with time, and have good
chemical resistance. Contemporary optimization principles (Section 3.3) requires
simultaneous penetration by stable particulate grouts to ensure good long-term
performance.

Also of considerable potential is the use of silacsols. Silacsols are solution grouts
formed by reaction between an activated silica liquor and a calcium-based inorganic
reagent. Unlike the sodium silicates discussed above - aqueous solutions of colloidal
silica particles dispersed in soda - the silica liquor is a true solution of activated
silica. The reaction products are calcium hydrosilicates with a crystalline structure
similar to that obtained by the hydration and setting of Portland cement, i.e., a
complex of permanently stable crystals. This reaction is not therefore an evolutive
gelation involving the formation of macromolecular aggregates, but is a direct
reaction on the molecular scale, free of synersis potential. This concept has been

employed in Europe since the mid-1980s (Bruce, 1988) with consistent success in fine-medium sands. The grout is stable, permanent, and environmentally compatible. Other important features, relative to silica gels of similar rheological properties, are:

- their far lower permeability;
- their far superior creep behavior of treated sands for grouts of similar strength (2 MPa);
- even if an unusually large pore space is encountered, or a large hydrofracture fissure is created, a permanent durable filling is assured.

Finally, the concept of "precipitation grouts," as addressed by Naudts (1996), may have major, if infrequent, application. Solutions are injected into the groundwater which trigger a chemical reaction with metal ions in the groundwater, producing a precipitation of durable crystals or complex metal agglomerations, which block flow paths.

Illustrative Case Histories

During the last few years, the authors have consulted on a large number of major projects involving the stopping of high velocity, high head flows under and into major structures. These are not problems that are unique to dams (quarries, mines, tunnels, and deep basements are equally susceptible), and they are problems that are encountered worldwide, as witnessed by recent projects in the Philippines, Argentina and Malaysia. Often such projects become highly political and sensitive such are the technical, commercial, public safety, and environmental and scheduling consequences they generate, and for this reason, clients are frequently loath to allow details of the work to be publicized. Though understandable, this approach does not help advance the state of knowledge in the industry since it denies access to extremely valuable case histories, frequently executed using innovative techniques and methods. Given also the space restrictions, this paper provides summary data from only three recent projects:

- Dworshak Dam, ID;
- Tims Ford Dam, TN; and
- Potash Mine, New Brunswick, Canada.

More detailed information on Dworshak Dam is provided by Smoak and Gularte (1998), and on Tims Ford Dam by Bruce et al., (1998).

Dworshak Dam, ID

Background

Dworshak Dam had been constructed for the Corps of Engineers on the North Fork of the Clearwater River, approximately 55 km east of Lewiston, ID by 1972. The dam has a structural height of 219m, is the highest straight axis concrete gravity dam in the Western Hemisphere, and the third highest dam in the U.S. The dam crest is 1002m long at elevation 492m. The dam provides flood control, power generation, fish migration, and recreation.

The bedrock under the left abutment is composed of competent granite gneiss with foliations dipping 15 to 30 generally to the west. Features that were nearly vertical and striking northeast to southwest are also present. Inspection of the foundation from the dam adits revealed very competent rock that is slightly to very slightly fractured and jointed with widely scattered shearing. The fractures/joints are commonly infilled with clay and mica.

Foundation permeability as determined by pressure testing in boreholes, was moderate to very low, progressively decreasing with depth from 1 x 10^{-3} cm/s (10-m depth) to 5 x 10^{-7} cm/s (75-m depth). During dam construction, a single line grout curtain was created from a basal grouting gallery using contemporary methods, and a drainage curtain was constructed downstream of this curtain from the same gallery.

The Problem

Seepage flows at full reservoir elevation from the left abutment drains were relatively constant until mid 1984 with flows from the drains in Monoliths 14 through 17 totaling about 2,300 l/min. After 1984, a significant increase in rate of flow began. Seepage by 1987 had increased to 4,500 l/min, and measurements taken in mid 1996 revealed a total flow of 9,500 to 11,500 l/min from the drains in Monoliths 14 through 17. More than half that total was coming from the drains located within Monoliths 16 and 17. The flow from the drains seemed to be clear but observation of the various collection flumes showed that fracture infill material was being eroded. Although foundation uplift pressures remained well below the original design assumptions, there was concern that if flows increased beyond drain capacity an increase in uplift pressures could occur. Such flows exceeded the capacity of the left abutment drainage gallery, overtopping stairs, landings and gallery walkways causing personnel safety concerns. In addition, vertical joint drains located between Monoliths 14/15, 15/16, 16/17 and 17/18 showed significant leakage into the gallery directly through their waterstops. This was possibly due to waterstop failure or improper installation, or inefficiencies in concrete placement.

Numerous investigations and evaluations of the problem water flows were performed during the period 1984 through early 1995. The overall conclusions from the investigations were that:

1. The flow was coming through fractures intercepted by the left abutment foundation drains in Monoliths 15 through 19. These fractures were interconnected and drain hole cross communication was common.

2. Some individual drains had flow as high as 750 l/min and pressures as high as 0.7 MPa. Most drain hole flows, however, were significantly less than 400 l/min and the pressures were below 0.3 MPa.

3. Vertical rock fractures may not have been intercepted and grouted during construction of the original grout curtain.

4. Some clay infill material from the foundation fractures was being piped into the grouting gallery.

The Solution

A remedial program was initiated between May and December 1997, with three main actions:

1. Installation of additional dam instrumentation to continuously monitor uplift pressure and leakage flows and the setup of an instrumentation database to manage the instrumentation data generated during and after remedial grouting.

2. Remedial foundation drilling and grouting in Monoliths 15 through 19 from the gallery with the pool reduced from elevation 488 to 457m.

3. Repair leaking upstream monolith joint waterstops by grouting, including reestablishing any grout sealed downstream monolith drains.

An innovative procurement system was used by the Corps of Engineers focusing heavily on the technical proposal of the contractor, Partnering, and full technical cooperation between all parties. The specifications contained the following requirements of the contractor:

1. Furnish and install additional and replacement uplift pressure instrumentation, crack/joint displacement meters, seepage flow monitoring instrumentation and open tube piezometers complete with a database monitoring system capable of presenting the instrumentation data in spreadsheet form.

2.   Remedial Foundation Grouting

    a.   Perform remedial grouting using 45 m³ of water reactive, fast setting solution grout, injected in the existing foundation drains, to construct a temporary downstream curtain.

    b.   Drill, and inject with cementitious grouts, a multi-row permanent upstream grout curtain including 6,000 lin. m of grout hole drilling.

    c.   Drill 2,500 lin. m of relief drain holes to establish a new row of downstream foundation pressure relief drains to replace those grouted during formation of the temporary grout curtain.

3.   Controlling Monolith Joint Leakage

    a.   Install packers in leaking monolith joint drains to reduce and/or control leakage to less than 38 l/min. (This work was specified because testing had verified that there was connection between the leaking relief drains and the foundation bedrock joints and fractures.)

    b.   After completion of curtain grouting, repair upstream leaking monolith joint waterstops by grouting the upstream vertical drain holes.

    c.   After grouting the upstream vertical monolith drains, reestablish the downstream monolith joint drains by cleaning, or by drilling replacement downstream joint drains.

Highlights of Construction

- The work was constructed under extremely difficult working conditions in the steeply dipping 1.8 x 2.4m gallery, inundated with cold seepage water.
- An extensive environmental protection program was successfully instituted.
- An extensive Contractor Quality Control Plan was successfully implemented.
- Data from instrumentation recording uplift pressure, hydraulic head, and gallery flow were monitored by laptop computer in the gallery, continuously during the work, and also after its completion.
- Sealing of the monolith joint drains (through which 50% of the flow was occurring) was effected by using the Multiple Packer Sleeve Pipe (MPSP) system (Bruce and Gallavresi, 1988), and polyurethane and modified cementitious grouts.
- The 63 existing NX drain holes (as deep as 67m) were first sealed using the MPSP system polyurethane and accelerated cementitious grouts. Remnant flows were minimal.

•   The 2-row remedial grout curtain (upstream of the original) was then installed
         to a depth of 40m using conventional stage grouting and modified
         cementitious grouts.
•   The replacement drainage curtain was then installed in the line of the original
         holes to a depth of 30m.

Effect of Treatment

         Total gallery flows had been reduced to less than 100 l/min by December
1997, following the drilling of 6,000 lin. m of grout and drain holes, and the injection
of over 45 m^3 of polyurethane and 170 tonnes of cement, within a 150-calendar day
schedule. No uplift pressures were recorded on the foundation. These observations
were made with the reservoir elevation at about Elevation 466m.

Tims Ford Dam, TN

         Background

         Tims Ford Dam is an embankment structure constructed on the Elk River
approximately 14 km west of Winchester, TN. This water regulating Tennessee
Valley Authority (TVA) structure is about 460m long with the crest at elevation
277.4m. The right (west) abutment of the dam is a ridge running nearly north-south
(Figure 1), and consisting of clay and weathered chert overburden overlying a karstic
foundation of various limestones. The crest of this right rim abutment varies in
elevation from 287m to about 292m with the top of rock generally around Elevation
274m. The maximum pool elevation is at Elevation 270.7m.

         The Problem

         In May and June 1971, two leaks designated Leaks 8 and 6 appeared on the
downstream side of the right rim during initial filling. Leak 8 was approximately
45m upstream of the dam base line. Exploratory drilling and dye testing were
performed along the right rim for a distance of 630m upstream of the dam baseline.
This work led to grouting a curtain line of holes using cement based grouts
containing calcium chloride accelerator to withstand the water flow velocity. At that
time, dye connection times from curtain to Leak 6 were recorded in the range of 4 to
8 hours. No attempt was made to seal it. The major outflow from Leak 6 emitted
from two vertical features at Elevation 260m, some 290m upstream of the dam
baseline, and formed an unnamed stream traveling approximately 1000m to the Elk
River. An outflow monitoring program was begun and data from that program
showed that the outflow varied directly with reservoir level. During the period 1971
through 1994, Leak 6 peak outflow volume slowly increased to about 15,000 l/min.
In 1994, however, following record drawdown of the reservoir, the Leak 6 outflow
volume increased dramatically in 1995 to over 29,000 l/min. TVA determined that

remedial grouting should be performed to reduce the Leak 6 outflows to less than 4,000 l/min at maximum pool.

An exploratory drilling program was performed during February to April 1997 to better define the existing foundation conditions and provide information necessary to design the remedial grout curtain. This program consisted of drilling a total of 20 vertical and inclined holes, permeability testing in stages, and dye testing to develop flow connection times and paths to Leak 6. The exploratory program provided the following conclusions:

1. Progressive erosion of collapsed and/or desiccated karstic feature infill material was the likely cause of the increased seepage. These features were controlled by solutioning along bedding planes and vertical or near vertical joint sets. Open features in excess of 6m deep were detected. Several dye test connection times of only minutes were encountered to the seep.

2. The bottom elevation of the remedial grout curtain as indicated by the geology and permeability, was estimated as Elevation 256m.

3. The southerly extent of the remedial grout curtain was geologically well defined.

4. The middle and north end of the exploratory area was less uniform with high water takes, cavities and open features, very fast dye connection times and the possibility of an undetected open channel to Leak 6. (The possibility of an open channel was reinforced by the occurrence of low permeability areas near the north end on either side of a high permeability area, thus leaving the location of the north end of the curtain somewhat questionable).

There was strong evidence that there would be substantial water flow through the features of the foundation rock during remedial grouting.

The Solution

A multirow remedial grout curtain was designed, approximately 240m long. The holes were inclined at 30 degrees to the vertical to encourage intersection of (sub)vertical features and were oriented in opposite directions in the two outside rows. Primary holes in each row were foreseen at 12-m centers, with conventional split spacing methods to be employed (to 3-m centers). The central, tightening, row was vertical. The grouting was to be executed between Elevations 270.7 and 256m - locally deeper if dictated by the stage permeability tests conducted prior to the grouting of each stage.

Because of the suspected high flow conditions, the downstream curtain row holes that encountered voids and active flow conditions were designated to be grouted with fast-setting (1 to 3 minute set time) hydrophillic polyurethane resin to provide an initial semi-permanent flow barrier. Holes that did not encounter voids or active flow were to be grouted with cementitious grouts. Upon completion of the downstream row it was anticipated that the active flow conditions would be mitigated, thus allowing the entire upstream row followed by the third, central, closure row to be grouted with cementitious grouts to form a permanent and durable grout curtain. The grouting was designed to be performed using upstage methods although it was anticipated that poor foundation conditions could locally require utilization of downstage methods. The grout holes were to be cased through the overburden from the surface to the top of the curtain. The Owner's goal was to reduce the peak seepage to about 4,000 l/min and to focus only on the major features (i.e., not to specifically treat the smaller fissures).

The Specifications contained provisions that required monitoring and limitations to outflow pH and turbidity to protect the downstream environment. TVA agreed to draw down the reservoir to Elevation 260.6m (3m below minimum normal pool) to minimize hydraulic gradient and flow through the rim. The curtain was to be constructed by first grouting the far ends, so conceptually channeling the flow through a middle zone which would then be grouted.

Highlights of Construction

- When drawdown of the reservoir reached Elevation 261.8m the outflow from Leak 6 completely and naturally stopped. As a consequence, much of the grouting work could be done in "no flow" conditions; therefore, largely eliminating the need for the polyurethane grouts, and extending the applicability of cement based formulations.
- Larger than anticipated open or clay-filled features were encountered especially in the upper 6m or so of the curtain. For technical, commercial, environmental and scheduling reasons, such features were treated with a low mobility "compaction grout" (slump 50 to 150mm; containing also water reducing and antiwashout agents).
- A suite of cement-based grouts were developed to permit the appropriate match of mix design and "thickening sequence" to the particular stage conditions as revealed by drilling and permeability testing (both multi- and single-pressure tests). Details of the initial mixes and their application are provided in Tables 2 and 3.

Ingredient	Unit	Mix A	Mix B	Mix C	Mix D
Water	lb	141	141	94	94
Bentonite	lb	4.7	9.4	4.7	4.7
Cement	lb	94	94	94	94
Rheobuild 2000	oz	15	30	20	30
Rheomac UW450	oz	0	0	0	5
Volume of batch	gal	20.8	21.0	15.1	15.1
Specific gravity		1.39	1.4	1.53	1.53
Bleed	%	<5	<1	<1	0
Kpf	min$^{-1/2}$	<0.104	<0.042	<0.042	<0.042
28-Day Compress.	psi	500	500	800	800
Marsh time	sec	35	50	60+	100+
Stiffening time	hh:mm	4:30	4:30	4:00	4:00
Hardening time	hh:mm	10:30	8:30	8:00	8:00
Water and slurry volumes					
Bentonite slurry volume	gal	8.0	16.1	8.0	8.0
Additional water volume	gal	9.9	2.8	4.2	4.2

Table 2. Compositions and properties of cement grout mixes,
Tims Ford Dam, TN

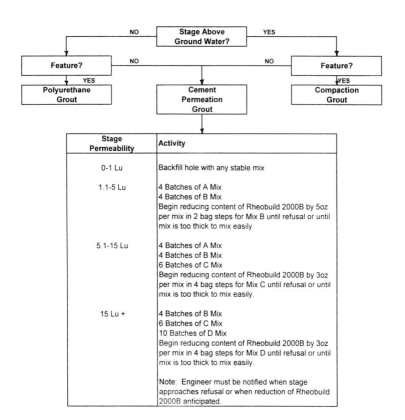

1.   Refusal will be defined as a flow of 1 gpm measured over a 10-minute period at the
      target pressure of 1 psi per foot of depth.
2.   No more than 60 batches of cement grout will be injected into a given stage on one
      12-hour shift.
3.   Compaction grout may be used for features below the water table in the future but,
      until such a decision is confirmed, only polyurethane will be used in such features.

Table 3. Flow chart providing guide to mix selection and variation, Tims Ford
Dam, TN.

• In response to conditions revealed during the treatment, observations of the seepage and further dye testing, extra groups of holes were added at the north end of the curtain, including 11 orthogonal to the original curtain, to allow specific treatment of key features.

• About 15,500 m³ of compaction grout, 1530 liters of polyurethane, and 605 m³ cement based grouts were injected into a total of 250 holes (comprising 3400 lin. m of rock drilling).

Effect of Treatment

Throughout the work, closest attention was paid in real time to data from the drilling, water testing, and grouting activities in addition to information from leak monitoring, piezometers and dye testing. The curtain was thus brought to an engineered refusal. During refilling of the reservoir, the leak was eliminated with the level at elevation 265m, when, for financial reasons, the work was terminated. The most recent reading, with the lake at elevation 269m indicates a seepage of around 950 l/min (net of surface runoff contributions) - about 5% of the flow at equivalent lake elevation prior to grouting. Data from piezometers and dye testing support the existence of an efficient curtain.

Potash Mine, New Brunswick, Canada

Background

During the late fall of 1996, minor leaks were detected in one of the highest areas of a major potash mine, near Sussex, N.B. This mine operates with the room and pillar method of excavation. In the area of the inflow, the back of the stopes was close to the shale caprock. At the time, the water inflow was judged insignificant as it did not affect production, and so was not treated, although an accelerated backfill program in this area was launched to provide more support and to try to prevent the problem from escalating. It was hoped that the seepage would drain a small isolated reservoir in the overlying strata and would eventually disappear.

However, the inflow continued to increase, as the roof started to deteriorate and collapse. Fresh water that enters a potash or salt mine is always a significant threat, since it can cause rapid solution. By late May 1997, the inflow had escalated, to a point that the mine was forced to shut down. Inflows were estimated to be in the order of 10,000 to 15,000 m³ per day. The water was fresh, and believed to originate predominantly from a water-bearing zone located approximately 200 to 300m above the mining horizon. The inflow dissolved thousands of tonnes of salt per day and cut a pathway down to the basalt below the salt horizons. From there, it moved laterally to a point where it was intersected and pumped away. However, the mine's dewatering system could only handle 5,000 m^3 per day, which resulted in a gradual flooding of the mine.

Solution

Following suspension of mining activities, the Owners selected the program proposed by ECO, even though it was understood that the chances of success were estimated at only 1 in 3, so severe was the structural deterioration caused by solutioning. The foreseen methodology featured the injection of hot bitumen in conjunction with modified cement based grouts, a long used concept that had been greatly refined and optimized in the course of more recent projects. Importantly this plan was to be implemented in conjunction with the simultaneous drilling of pressure relief holes, installed from the underground workings, to control the inflow and channel it to pump stations. These holes would also serve to provide data on the effectiveness of the grouting operation in real time. If pressure relief were not properly effected, then rapid build up of water pressure in the cavern and formation would otherwise lead to hydrofracturing of the formation, and so increased flow rates.

Two inclined drill holes were to be advanced from the surface to the cavern deliver the substantial amounts of materials: one line for bitumen, the other for cement grouts. The cavern was located 700m below the ground surface.

Highlights of Construction

• Directional drilling was used to successfully drill the two nearly vertical but curved holes in the cavern.
• Dye and air tests were performed through these holes to verify connection to the inflow, establish the size of the rubble pile at the base of the cavern, and calculate the volume of the cavern (approximately 19,000 m^3).
• Injection of hot bitumen had never before been attempted to such depth, and the installation included grouting of the lower casing with insulting cementitious grout, hot oil circulation concentric piping systems, thermal expansion joints, bitumen delivery pipe with stringer and rupture discs, two thermocouples and wellhead attachment, bitumen reheating systems and heated storage tanks, and hot oil heating systems.
• For operational reasons, only two pressure relief holes had been completed prior to the grouting operation commencing.
• The bitumen plant was constructed to provide an average capacity of 20 m^3/hr without interruption to handle the foreseen volume of 6,000 m^3. The hot oil system was required for preheating the bitumen line to 125 C, as was the passage of a limited volume of "soft bitumen". Bottom hole temperatures exceeded 150 C before the "hard bitumen" could be injected.

- Six different modified cementitious grout formulations were used for void filling and formation grouting activities. These mixes had well defined performance characteristics (antiwashout, low pressure filtration coefficient, no bleed, high strength, durable, high abrasion and erosion resistance) within a wide range of viscosities and specific gravities. The antiwashout additive was added, for logistical reasons, downstream of the mixer.
- A fully automated and computerized colloidal mixing and pumping plant, capable of producing 60 m³/hr of grout was specially developed. Continuous QA testing of grout properties was executed by the supervisory staff.
- An intensive manual and electronic monitoring program was implemented, with computers at the bitumen site, the cement site, and the main control center recording dozens of variables in real time on grouting progress, and the response of the groundwater.

Effect of Treatment

The mechanical execution of this enormous and difficult task was flawless. After three days of continuous injection, following a detailed program a combined total of 2,000 m³ of bitumen and cement grout had been successfully injected. The inflow began to decrease within 24 hours and the formation pressure began to rise. By the end of the third day, the inflow was completely stopped and the formation pressure continued to rise. Grouting continued at the same injection rates (25 m³ of bitumen per hour and 45 m³ of cement based grouts per hour). Within 36 hours, there was no more washout of the cement based grout.

On Day 5, however, a major collapse and settlement of the rubble pile and eroded salt backfill took place triggered by the greatly increased hydrostatic pressure. Although this event was predicted and special measures had been taken underground for the occurrence of this event, the devastation caused by the resulting "tidal wave" was overwhelming. After generating an inflow rate of over 3,500 m³/hr until the cavern had emptied itself, it returned to the pre-grouting flow rates within about 3 hours.

The grouting continued at slightly increased rates from both holes. Towards the end of Day 7, the rate of inflow started to decrease and the formation water pressures started to rise again. The increase of formation water pressure with time was much slower than during the first operation, indicative of a much larger cavern, caused by the collapse during Day 5. Towards the end of Day 10, the leak had again been reduced to a trickle and formation pressures were recovering faster. The inflow rates fluctuated for a few days: each slight increase in inflow triggered a decrease in formation pressure and vice versa.

Suddenly, during the thirteenth day of grouting, the entire area around the cavern collapsed. Most likely the undercutting, by solutioning of the salt layers at or near the contact with the basalt had been too extensive. A large block of ground collapsed, followed by a tidal wave, which flooded thousands of cubic meters of water into the mine from the cavern in 5 hours. A last effort was made involving the injection of bitumen at pump rates of 40 m^3 per hour and cement grout in conjunction with sodium silicate (via 2 concentric pipes) at a rate of almost 60 m^3. However, the new cavern had become so large that the consultants, owners, and management all came independently to the same sad conclusion; the undermining by the fresh water had caused so much damage that the mine could not be salvaged, under these conditions.

So, after almost 15 days of continuous grouting, totally without down-time, the operation was terminated. A combined total of over 22,000 m^3 of bitumen and cement grout had been injected during this period.

Final Observations

These three case histories have many elements in common:

1.   The advantage of having access to accurate historical records.
2.   The necessity of careful research and exploration towards determining the nature and extent of the problem and so allowing engineered design of the solution.
3.   The need to select efficient, knowledgeable, experienced, and committed specialists, as both consultants and contractors.
4.   The need to select appropriate materials, equipment, and methods, and the possession of a fundamental level of understanding to modify these appropriately in the light of actual conditions on site ("Responsive Integration" - Bruce et al., 1993).
5.   The need for real time monitoring and analysis of drilling and grouting data.
6.   The need for the highest levels of QA/QC on materials and mixes.
7.   The needs to establish appropriately quantified and measured " measures of success", and to "baseline" these prior to commencing the treatment.
8.   The benefits of using contemporary cement grout admixtures.

Such works are typically conducted under adverse geological, site and logistical conditions and considerable financial, environmental and time pressures. However, these case histories illustrate quite clearly what can be achieved, assuming that the eight elements listed above are properly observed.

Acknowledgments

The authors wish to express their full appreciation for the input from their colleagues in the ECO Group, and the cooperation and camaraderie of their counterparts amongst the various agencies, consultants, contractors, and suppliers involved. You judge people in adversity.

Units

In this paper, the following "soft conversions" have been used:

1m	≡	3.3 ft
1 liter	≡	0.26 gallons
1 m³	≡	35.3 cf
1 MPa	≡	145 psi

References

1. AFTES, (1991). "Recommendations on Grouting for Underground Works," *Tunneling and Underground Space Technology*, Vol. 6, No. 4, pp. 383-461.

2. Bruce, D.A. (1988). "Developments in Geotechnical Construction Processes for Urban Engineering." Civil Engineering Practice 3 (1) Spring, pp. 49-97.

3. Bruce, D.A. (1990a). "Major Dam Rehabilitation by Specialist Geotechnical Construction Techniques: A State of Practice Review." Proc. Canadian Dam Safety Association 2nd Annual Conference, Toronto, Ontario, September 18-20, 63 p. Also reprinted as Volume 57 of the Institute for Engineering Research, Foundation Kollbrunner-Rodio, Zurich, September.

4. Bruce, D.A. (1990b). "The Practice and Potential of Grouting in Major Dam Rehabilitation." ASCE Annual Civil Engineering Convention, San Francisco, CA, November 5-8, Session T13, 41 pp.

5. Bruce, D.A. (1992). "Drilling and Grouting Techniques for Dam Rehabilitation." Proc. ASDSO 9th Annual Conference, Baltimore, MD, September 13-16, pp. 85-96.

6. Bruce, D.A. and F. Gallavresi. (1988). "The MPSP System: A new method of grouting difficult rock formations." ASCE Geotechnical Special Publication No. 14, "Geotechnical Aspects of Karst Terrains, " pp. 97-114. Presented at ASCE National Convention, Nashville, TN. May 10-11.

7. Bruce, D.A., Luttrell, E.C. and Starnes, L.J. (1993). "Remedial Grouting using Responsive Integration^{sm}." Proc. ASDSO 10th Annual Conference, Kansas City, MO, September 26-28, 13 pp. Also in Ground Engineering 27 (3), April, pp 23-29.

8. Bruce, D.A. Littlejohn, G.S., and A.M.C. Naudts (1997) "Grouting Materials for Ground Treatment: A Practitioner's Guide", *Grouting -*

Compaction, Remediation, Testing, ASCE, Geotechnical Special Publication No. 66, Ed. by C. Vipulanandan, pp. 306-334.

9. Bruce, D.A., J.A. Hamby, and J. Henry. (1998). "Remedial Foundation Grouting at Tims Ford Dam, TN." Paper in preparation for 15th Annual Conference on Association of State Dam Safety Officials, Las Vegas, NV, September, 1998.

10. Gause, C. and D.A. Bruce, (1997). "Control of Fluid Properties of Particulate Grouts (1) General Concepts."*Grouting - Compaction, Remediation, Testing*, ASCE, Geotechnical Special Publication No. 66, Ed. by C. Vipulanandan, pp. 212-229.

11. Naudts, A.M.C., (1996). in Brown, Robert Wade, "Practical Foundation Engineering Handbook," McGraw-Hill, New York, NY. pp. 5.277-5.400.

12. Smoak, W.G., and F.B. Gularte. (1998). "Remedial Grouting at Dworshak Dam." This conference.

SELECTION OF CEMENT-BASED GROUTS FOR SOIL TREATMENT

Maria Caterina Santagata[1] Student Member, ASCE and Mario Collepardi[2]

ABSTRACT

The design of high performance grouts for soil treatment requires consideration of many aspects of the grouting mixture including the selection of the binder (composition and fineness), the admixture (e.g. type and dosage of superplasticizer), and the water-binder ratio.

This paper describes the results of a laboratory program conducted to establish criteria for cement-based grout design. Grouts were manufactured employing three pozzolanic microcements, identical in composition, but different in fineness (D_{98} = 9, 17, 20 μm) and an acrylic-based superplasticizing admixture. Grout injectability was evaluated through laboratory injection tests performed on columns of four different sands with hydraulic conductivity ranging from 2.1×10^{-4} to 1.2×10^{-3} m/s. Compression tests on specimens of the grouted sand indicated the effectiveness of the treatment. In addition, mercury intrusion porosimetry was employed to characterize the microstructure of the impregnated soil.

Results indicate that by minimizing the water-cement ratio of the grout the effectiveness of the treatment, in terms of higher strength and smaller void size of the grouted medium, is greatly enhanced. The minimum water-cement ratio that guarantees successful permeation of the soil varies greatly depending on the properties of the porous medium (permeability) and of the grout (fineness of the binder, use of superplasticizing admixture). Curves relating the composition of the grout to the permeability of the untreated soil and to the desired properties (strength and void size) of the treated medium were also obtained. These curves may be used to select the most appropriate grout for a given soil, based on the required strength of the grouted medium.

---

[1] Research Associate, Department of Earth and Material Science, University of Ancona, Ancona, Italy; Research Assistant, Department of Civil and Environmental Engineering, Massachusetts Institute of Technology, Cambridge, MA.
[2] Full Professor, Department of Earth and Material Science, University of Ancona, Ancona, Italy.

INTRODUCTION

As grouting continues to be employed in applications requiring superior performance of the treated medium, in terms of increased strength and durability and reduced permeability [e.g. 1, 2], the need for rational criteria for the design of these mixtures has become a necessity.

In previous work [3], the authors have examined the effect of various mixture parameters, in particular the composition and the fineness of the binder and the type and dosage of superplasticizer, on the performance of grouts, in terms of groutability in laboratory sand specimens, and the compressive strength of treated soils.

The scope of the present work is to investigate how the medium to be treated affects the selection of the grouting mixture and the effectiveness of the treatment in terms of increased strength and reduced porosity.

Columns of four different granular media, with permeability ranging between 2×10^{-4} and 10^{-3} m/s, were considered. One Portland microcement blended with 36% natural pozzolan, available in three fineness classes [4] ($D_{98} = 9$, 17, and 20 μm) and one superplasticizing admixture were used for the grouts. More than 30 injection tests were performed. In addition, mercury intrusion porosimetry (MIP) was employed to investigate the microstructure of the grouted sand.

A limited number of measurements of the Marsh viscosity, the compressive strength and the stability of the grouts were also conducted.

The work presented here was performed as part of a broader research program ongoing at the University of Ancona, aimed at developing a rational framework for the design of cementitious grouts for treatment of porous systems (e.g. sand deposits, masonry walls). Current work involves the assessment of the durability of grouts in a variety of aggressive environments, and, through the use of a rotational rheometer, the investigation of the factors affecting their rheology.

MATERIALS

Microcements

Three industrial blended microcements of identical chemical composition but different fineness were employed in the testing program. The three microcements contained 60% portland clinker (with $C_3A = 9.5\%$), 36% natural pozzolan, and 4% natural anhydrite as a set regulator. Table 1 summarizes the composition of these raw materials.

	SiO_2	Al_2O_3	Fe_2O_3	CaO	MgO	SO_3	K_2O	Na_2O	Cl	L.o.i.	Tot.
Clinker	22.47	5.07	1.94	65.49	2.11	1.4	0.88	0.15	0.01	0.22	99.74
Anhydrite	3.12	0.95	0.33	36.09	2.56	49.6	0.26	0.09	0.04	6.57	99.61
Pozzolan	61.68	18.10	7.91	3.37	3.09	0.0	3.01	1.24	0.03	1.20	99.63

Table 1. Chemical Composition (% by mass) of Raw Materials

The grain size distributions of the three binders were determined using a laser granulometer and are shown in Figure 1.

Throughout this paper, and consistently with other work published by the authors [3], these microcements will be referred to as *P-Z-9*, *P-Z-17* and *P-Z-20*, where the first letter (*P*) indicates the type of clinker, the second (*Z*) the type of mineral addition, and the number corresponds to the D_{98}, the characteristic diameter, expressed in microns, corresponding to 98% finer material, which is used to quantify the fineness of the binder [4]. The lower the D_{98}, the finer the microcement. For the three binders considered, D_{98} is equal to 9, 17 and 20 μm respectively.

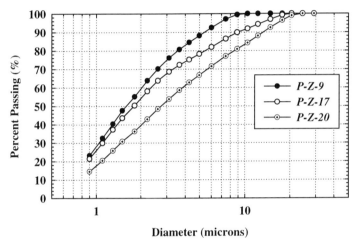

Figure 1. Grain Size Distributions of the Three Microcements

Superplasticizer

An acrylic based superplasticizer (AP) was employed for the majority of the mixtures presented in this paper. This product is a 30% aqueous solution of carboxylated acrylic ester polymer. Previous work [3] indicates that, compared to the more commonly employed melamine and naphthalene based admixtures, this superplasticizer is much more effective in enhancing the fluidity and improving the injectability of a grout, as well as reducing the fluidity loss.

Various dosages of this superplasticizer were considered in the mixtures. All are expressed in terms of percentage of active polymer by mass of cement.

Sands

Columns of sand for the injection tests were manufactured according to the procedure described in the following section employing four different natural soils referred to in the following as coarse Po sand, Ticino sand, fine Po sand, and Musone sand. Grain size distributions for these four soils determined by sieve analyses are shown in Figure 2. Table 2 summarizes a few parameters for these sands, namely the coefficients of uniformity and the values of D_{10} and D_{15} that are used in practice to define injectability coefficients [5]. Note that the Ticino sand and the Musone sand are both characterized by approximately the same value of D_{15}, although their overall grain size distributions are quite different.

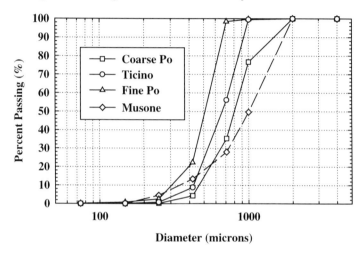

Figure 2. Grain Size Distributions of Four Sands

	Coarse Po	Ticino	Fine Po	Musone
Cu	1.87	1.71	1.83	3.53
D_{10} (mm)	461	427	306	348
D_{15} (mm)	502	450	346	448
k ($x10^{-4}$m/s) #tests	11.76$\pm$1.27 (9)	7.82$\pm$0.66 (14)	4.89$\pm$0.22 (5)	2.09$\pm$0.35 (5)

Table 2. Characteristics of Four Sands

METHODS

Figure 3 shows a schematic view of the apparatus employed to perform the injection tests described in this paper. It consists of three basic components connected by tubing and valves: the air compressor, the tank containing the grout, and the sand column to be injected.

The sand columns, approximately 46 cm tall with a 3.8 cm diameter, were manufactured by pouring and then compacting the sand inside a tube made from PVC sheet. A gravel filter was placed both below and above the sand sample.

Complete details of the procedures employed to prepare the columns are provided in [6].

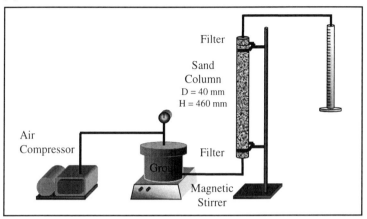

Figure 3. Schematic View of Injection Apparatus

        Prior to performing the injection tests the columns of sand were saturated with water and their permeability was measured under constant head. The average permeability values for the four sands with the respective standard deviations are summarized in Table 2. These measurements also provided an indication of the reproducibility of the characteristics of the sand columns. Grain size analyses and water content measurements at various heights along the sample were performed in the initial phase of the testing program to assure that the preparation procedures produced a homogeneous sample.

        The grouts were manufactured employing a rotary type laboratory mixer with the following standard procedure: the superplasticizer was mixed with the water; the water was added to the binder; the grout was mixed at low speed (150 rpm) for 2 minutes, left to rest for one minute, and then mixed at intermediate speed (180 rpm) for one additional minute. The grouts were injected from the base of the sand column by applying an initial pressure of 0.5 bar that was increased to maintain a constant outflow of water. The injection was halted either after the grout appeared at the upper filter or when the pressure reached 2 bars. The height of grout penetration inside the sand, which provided a measure of the penetrability of the grout in that medium, was measured once the injection was completed.

        The columns of grouted sand were left to cure inside the PVC casing in a vertical position at room temperature for thirteen days. At this time the PVC casing was removed and the grouted sand column was cut into 4 cm tall specimens using a water cooled rock cutting saw. The specimens were broken in unconfined compression the following day in a load frame equipped with a load cell.

Mercury intrusion porosimetry (MIP) analyses were performed on the grouted sand using a series 2000 Carlo Erba apparatus which has a capacity of 2000 bars, which permitted determination of the distribution of the voids having radius in the range of 30-200000Å. In this technique a dilatometer containing the sample is filled with mercury while under vacuum. As incremental pressure is applied to the mercury, the volume change is recorded. Since mercury is a liquid that does not wet most solid surfaces the size of the pore into which it can penetrate is inversely related to the applied pressure. Under the assumption of cylindrical pores [7], Washburn's equation $r = -[2\gamma(\cos\theta)]/P$ is used to relate the applied pressure (P) to the pore size (r) once the surface tension of mercury (γ) and the contact angle between the mercury and the pore wall (θ) are known[1]. Limitations and possible sources of error in this technique are discussed in [8]. As the dilatometer was only 15mm in diameter, the specimens tested were actually quite small and thus had to be selected carefully. To avoid error, three independent measurements were performed at any desired location. Note that, due to practical constraints, the MIP analyses were not performed at the same time as the compression tests, but approximately 2 weeks later (~1 month after injection). While this test is used primarily for research purposes, it provides a good illustration of the effects of the grout water-cement ratio on the void distribution in the grouted sand, which will ultimately affect its permeability.

Finally a limited number of bleed tests, Marsh viscosity (see [3] for description of procedure), and compressive strength measurements were performed on grouts manufactured with the three binders.

RESULTS

Marsh Viscosity and Compressive Strength of Grouts

Marsh cone tests were performed on three plain mixtures with w/c of 0.90 manufactured with each of the three binders. Measurements of the Marsh viscosity were performed immediately after mixing and then at regular intervals until the mixtures displayed infinite Marsh viscosity. The results for these tests, shown in Figure 4, indicate that the fineness of the microcement has a significant impact both on the initial fluidity and on the loss of fluidity with time.

Measures of the compressive strength at 1, 3, 7 and 28 days were performed on prismatic specimens cast from plain mixtures with w/c equal to 0.72. These results indicate that, at the early ages, the microcement fineness has no impact on the compressive strength (strength ~14 MPa for all three grouts). At 28 days however, the *P-Z-20* based grout does display marginally higher strength (30 MPa versus 27 MPa and 25.5 MPa respectively for the P-Z-17 and P-Z-9 mixtures). Most likely the relatively low water cement ratio employed, and the lack of vibration, did not permit complete elimination of the entrapped air in the mixtures containing the finer microcements.

---

[1] The following assumptions were made: γ=480 erg/cm^2, θ=141.3°.

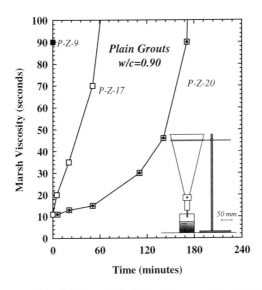

Figure 4. Marsh Viscosity for Plain Mixtures with w/c=0.90

Injection Tests in Sand

Tests were performed to evaluate the minimum water cement ratio that could be employed to obtain complete penetration of the sand columns, as a function of the sand and binder used. Both plain and superplasticized mixtures (with varying dosages of the acrylic polymer) were employed. This paper focuses primarily on the results obtained using the highest dosage of superplasticizer considered, which is 1.2% of pure product per mass of cement. Experience with this product indicates that further increase of the dosage beyond 1.2% yields negligible advantages [3]. The results are presented in terms of the 14 day-compressive strength of the grouted sand as a function of elevation from the base of the sand column.

The first set of results for coarse Po sand is presented in Figure 5a. The three curves of 14-day compressive strength versus elevation shown in the figure pertain to injections performed with grouts manufactured with each of the microcements considered, all with 1.2%AP. For each binder, the minimum w/c able to guarantee complete permeation of the column of coarse Po sand was determined by trial and error. As seen in Figure 5a, due to its higher surface area, the finest microcement (*P-Z-9*) required the most water (w/c=0.75) to achieve complete injection of the sand column. For the *P-Z-9* and *P-Z-17* based grouts, approximately constant values of the compressive strength were measured over the entire height of the column. This result reflects a fully successful injection. The average compressive strength was approximately 14 and 29 MPa respectively for

(a)

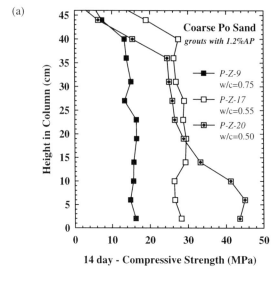

(b)

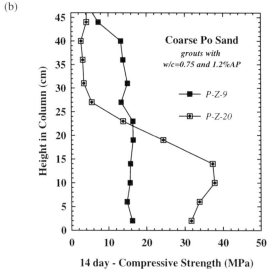

Figure 5. Compressive Strength of Grouted Coarse Po Sand

the *P-Z-9* grout with w/c of 0.75, and the *P-Z-17* grout with w/c equal to 0.55. The third plot, pertaining to the grout with the *P-Z-20* binder, shows a different trend of compressive strength versus height. Over the lower 10 cm of the column the strength measured is significantly higher (~40-45 MPa) than that measured on the other columns, thanks to the low w/c employed (0.50). However, a rapid decrease in strength is observed, and over a large portion of the column, the strength is low and the treatment is less effective than that obtained with the *P-Z-17* grout. This behavior is undesirable as it leads to unpredictable strength throughout the injected region.

The poor performance of the *P-Z-20* grout cannot be ascribed to excessive bleed (the total bleeded water was about 1% by volume of grout) or, as shown in Figure 5b, to insufficient fluidity. Here the results obtained by injecting columns of coarse Po sand with grouts manufactured with the *P-Z-9* and the *P-Z-20* binders, with w/c equal to 0.75 and 1.2% of AP are compared. Although much more fluid (see Figure 2), the mixture manufactured with the *P-Z-20* binder displays effective improvement of the properties of the medium only on the bottom 15-20 cm of the sand column. Above this height the strength decreases significantly and is much lower than that measured on the column grouted with the *P-Z-9* mixture. This can be attributed to the relatively large particles present in the *P-Z-20* microcement which limit its penetration into the pore space of the sand and contribute to create a plug at the bottom of the column thereby reducing further penetration of even the smaller particles of the binder. As a consequence, the grout that reaches the top of the column is increasingly poor in microcement content and unable to significantly improve the strength of the sand.

Plots similar to those presented in Figure 5a for coarse Po sand are shown in Figures 6-8 for Ticino Sand, fine Po sand and Musone sand respectively.

The results for Ticino sand (Figure 6) are qualitatively similar to those for coarse Po sand. The water cement ratio for complete permeation increases with the fineness of the binder from 0.60 for the *P-Z-20* grout to 0.85 for the mixture manufactured with the finest of the three microcements. A fairly uniform treatment of the sand over the majority of the height is obtained using the *P-Z-9* and *P-Z-17* based grouts. For the two finer microcements a lower w/c translates into a higher strength over the entire height of the column. The sand injected with the *P-Z-20* based grout shows a marked decrease of strength over the height of the column. Note that the columns grouted with the *P-Z-9* and *P-Z-17* based grouts show a decrease in compressive strength over the top 15 cm. This is due to insufficient fluidity of the two grouts. Unfortunately, practical constraints did not allow these injections to be repeated employing the slightly higher values of the water-cement ratio necessary to achieve uniform strength over the entire height of the sample.

For the two finer sands (Figures 7 and 8), only two plots are shown, as no *P-Z-20* based grout could be injected in these soils independently of the w/c employed. Note that in both cases the injectability criteria proposed by Mitchell [5]

would predict a successful injection. In the fine Po sand and the Musone sand the
values of w/c required to achieve complete penetration are significantly increased

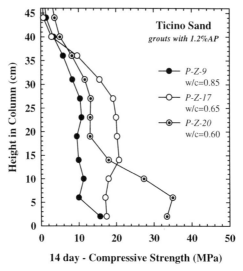

Figure 6. Compressive Strength of Grouted Ticino Sand

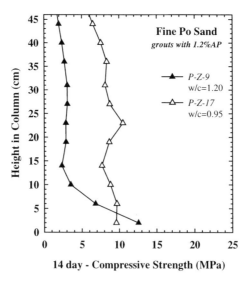

Figure 7. Compressive Strength of Grouted Fine Po Sand

with respect to the two coarser sands discussed. In all cases a fairly uniform
strength is measured over most of the sand column. However, except for one
instance (*P-Z-17* grout in fine Po sand), higher values of compressive strength are
measured at the bottom of the column. Note that this effect is more pronounced in
the case of the *P-Z-9* grouts. This result can be explained by the relatively high w/c
of these grouts. Despite the fact that the bleed results indicate that both the *P-Z-9*
mixtures are quite stable (~1% bleeded water by volume of grout), the higher w/c
may have resulted in the development of a pressure filtration mechanism through
the fine "filters" created by the two sands.

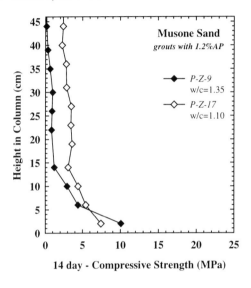

Figure 8. Compressive Strength of Grouted Musone Sand

     The results presented so far refer to mixtures manufactured with a relatively
high dosage of superplasticizer. With lower dosages of or no superplasticizer,
significantly higher water cement ratios had to be adopted to obtain equally
injectable grouts. For example, in the case of Ticino sand, as shown in Figure 8, it
was found that the w/c of the *P-Z-17* based grout had to be increased from 0.65
(1.2% AP) to 0.75, 1.0, and 1.4 respectively when 0.8, 0.4 and 0% of AP were
employed. Clearly the increased w/c translated into a less effective improvement of
the properties of the sand. In this particular case the average strength decreased
from about 15 MPa (grout with 1.2%AP), to 11 and 7 MPa when the AP dosage
was reduced to 0.8% and 0.4%, and to less than 4 MPa when no superplasticizer
was used. Further discussion on the effectiveness of superplasticizers as water
reducers in microcement grouts is provided in [3].

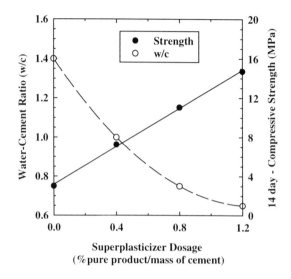

Figure 9. Relationship between Superplasticizer Dosage, w/c Necessary for Injection, and Grouted Sand Compressive Strength (Ticino Sand, *P-Z-17*)

The results for all the *P-Z-17* superplasticized (1.2%AP) grouts presented above are shown in Figure 10. The figure highlights how the maximum obtainable compressive strength is a function of the porous medium to be grouted, as increasingly high w/c grouts are required to permeate sands with lower permeability. Identical considerations apply to the *P-Z-9* based grouts.

The values of the minimum w/c necessary for complete permeation of the grout inside the sand column are plotted in Figure 13a as a function of the sand's permeability. Three curves, one for each of the three microcements, are shown. Only two results are shown for binder *P-Z-20* as, with this binder, the two finer granular media could not be permeated regardless of the water cement ratio employed.

For clarity only the data for grouts with 1.2% AP are shown, but additional curves for different dosages of AP superplasticizer could be drawn. Note that as the dosage of the superplasticizer decreases these curves would shift to the right, i.e. higher water cement ratios would be required to maintain equivalent injectability.

The dashed line in the figure defines limits of groutability that derive from both rheological and geometrical constraints. For combinations of values of k_{soil} and w/c included in this region complete injection of the sample is not possible either because the grout is not sufficiently fluid, or because the dimension of the binder particles is too large. While the geometric limits are expected to apply regardless of the type of microcement and the type and dosage of superplasticizer, the rheological limits are specific for the composition of this microcement and the 1.2% dosage of the AP superplasticizer.

Details on the binder and superplasticizer related factors affecting the fluidity of microcement grouts are provided in [3].

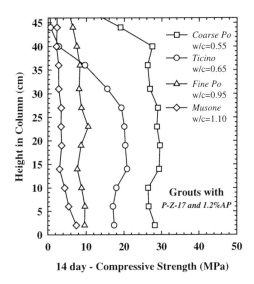

Figure 10. Effect of Sand on Maximum Compressive Strength (1.2%AP)

The improvement in the compressive strength of the sands after grouting varied greatly in the experiments performed. The average strength over the height of the column can be used as a comparative measure of the effectiveness of the treatment.

Values of the compressive strength are plotted in Figure 11 versus the water cement ratio of the mixture injected. The plot includes data for the *P-Z-9* and *P-Z-17* mixtures presented above along with data from additional experiments conducted with no or a reduced dosage (0.4 and 0.8%) of superplasticizer. All the data fall on one curve indicating that the compressive strength of the grouted sand is determined uniquely by the water cement ratio of the mixture independently of the type of sand, the fineness of the microcement, or the dosage of the acrylic superplasticizer. The shape of the curve is similar to that linking w/c and strength for ordinary concrete.

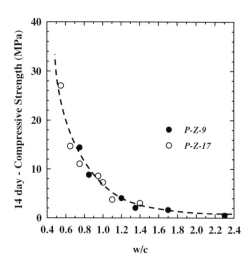

Figure 11. Effect of Grout Water-Cement Ratio on Average Compressive
Strength of Grouted Sands

Mercury Intrusion Porosimetry of Grouted Sand

MIP analyses were performed on fragments of grouted sand taken at mid-height from all the sand columns injected with mixtures manufactured with the two finer microcements and 1.2% AP [9]. Figure 12 shows the results obtained for one particular case (coarse Po sand injected with *P-Z-9* grout with w/c=0.75 and 1.2%AP). The first graph shown in the figure is the porogram, the plot of intruded volume versus pressure. This curve is converted into the pore volume distribution shown in Figure 12b by applying a cylindrical pore model as described above. From this curve the pore size frequency distribution shown in Figure 12c is then obtained. In this figure the size of the average radius corresponding to the steepest slope in the pore volume distribution curve (Figure 12b) is also shown.

Increasing the w/c of the grout causes the pore size distribution to shift to the right to higher values of the pore radius, and causes the average radius to increase. The values of the average radius determined for all of the tests performed are plotted in Figure 13c versus the water-cement ratio of the grout used to permeate the sand. In this figure the curve representing the sands grouted with the *P-Z-9* binder lies above that for the *P-Z-17* microcement, indicating that, at the same water-cement ratio, the more fluid mixture manufactured with the *P-Z-17* binder (see Marsh Viscosity curves in Figure 4) permeates the voids between the sand particles more effectively.

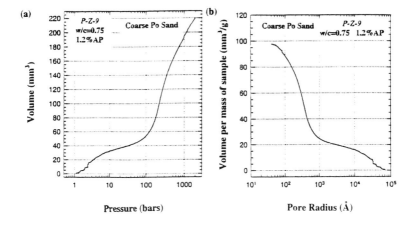

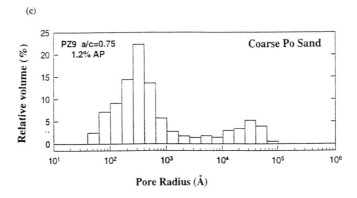

Figure 12. Results of Mercury Intrusion Porosimetry on Coarse Po Sand Grouted
with a *P-Z-9* Mixture with w/c=0.75 and 1.2% AP.

DISCUSSION AND CONCLUSIONS

Tests were performed to evaluate, as a function of the sand and microcement used, the conditions for complete injection of laboratory sand columns. The ability of a grout to penetrate into the pores of a deposit was found to depend on two factors: the size of the binder particles and the rheological properties of the mixture.

Existing injectability criteria [5, 10] consider only the first of these two aspects, as they are based on purely geometric considerations that relate the size of the solids of the grout to the size of the granular medium to be treated. It appears that for the grouts with high solid content considered in this study, these criteria do not always provide valid predictions. Table 3 summarizes the values of N (= $D_{15soil}/D_{85binder}$) and N_C (= $D_{10soil}/D_{95binder}$), the two groutability ratios discussed by Mitchell [5], for the three microcements and the four sands examined in this work. It can be seen that in all cases these ratios attain values significantly higher than the thresholds (24 and 11, respectively) above which, according to Mitchell, grouting should be consistently possible. In particular both groutability ratios would lead to expect successful injection of P-Z-20 based grouts in all four sands. In this work, however, it was found that, with this binder, injection of the sand columns was at best difficult, and in two cases (Fine Po Sand and Musone Sand - shaded cells in Table 3) impossible. In these two cases also the groutability limits based on the permeability of the porous medium ($D \leq C \cdot K^{1/2}$) examined by De Paoli et al. [10] are not verified, regardless of the value of D (D_{50}, D_{85}, D_{95}, etc.) used. Also note that for the two sands (Ticino and Musone) characterized by very similar D_{15}, the same groutability ratio N is obtained. This does not reflect the great differences in the ease of grouting encountered while injecting these two porous media.

	P-Z-9		P-Z-17		P-Z-20	
	N	N_C	N	N_C	N	N_C
Coarse Po	120	68	72	38	48	28
Ticino	107	63	64	35	43	26
Fine Po	82	45	49	25	33	19
Musone	107	51	64	29	43	21

Mitchell (1981): N>24: grouting consistently possible; N<11: grouting not possible
N_C>11: grouting consistently possible; N_C<6: grouting not possible

Table 3. Groutability Ratios for Sands and Microcements Considered

In traditional grouts, characterized by high w/c values, the rheology of the mixture is rarely a critical factor in determining its injectability. However, as grouts are employed for higher performance applications, the need to reduce the w/c necessitates an understanding of rheology-based groutability limits. Through injection tests in four different sands with three microcements, groutability limits that take into account both the geometric and the rheological constraints were derived for the specific case of grouts containing 1.2% of an acrylic superplasticizer. They are presented in the form of a graph as shown in Figure 13a.

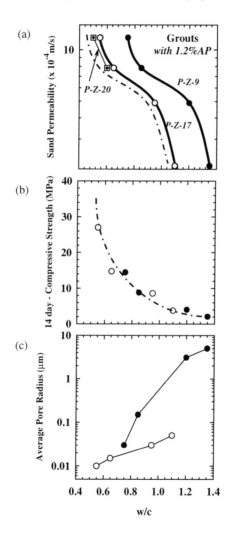

Figure 13. Curves for Grout Selection

From compressive strength tests and MIP analyses on the grouted sands, curves relating the composition of a grout (w/c, microcement, superplasticizer) to the permeability of the soil to be grouted and the properties (strength and size of pores) of the treated medium (Figure 13) were also developed. The water cement ratio of the grout was found to be the most important parameter governing the effectiveness of a treatment; and a unique relationship between compressive strength and w/c was found. Quite clearly the permeability of the deposit and the use of a superplasticizer come into play indirectly as they determine the minimum water content that can be used, and the ease of injection. The types of curves shown in Figure 13 may guide selection of the most appropriate mixture based on specific design requirements and field conditions. Although this study has mainly addressed the strength and, on a secondary basis, the porosity of the treated medium, equivalent correlations could be developed between the composition of the mixture and the permeability of the grouted medium.

While the picture presented above is valid for the limited set of conditions examined, this paper suggests a general framework for the selection of grouts used for soil treatment. When developing the mix design for a grout other aspects may require consideration, including the loss of fluidity, the heat of hydration, the volume stability, the durability to the environment, the bleed, the resistance to pressure filtration etc. Although experience in traditional concrete technology may certainly aid in solving many of these issues, the novelty of some of the materials used, the variability in the field conditions and the specificity of some of the environments encountered may often necessitate novel criteria.

ACKNOWLEDGEMENTS

The authors wish to acknowledge the assistance of Mr. Salvatore Fiore who performed many of the tests presented in this paper as part of the research for his degree in Civil Engineering at the University of Ancona. The assistance of the companies Gruppo Cementi Rossi of Piacenza, Italy and MAPEI of Milan, Italy that provided the materials is also greatly appreciated.

REFERENCES

[1.] Ahrens, E. H. (1997), "A New and Superior Ultrafine Cementitious Grout", *Grouting: Compaction, Remediation and Testing*, Geotechnical Special Publication No.66, Ed. C. Vipulanandan, pp. 188-196.

[2.] Berry, R.M. and Narduzzo, L. (1997), "Radioactive Waste Trench Grouting; A Case Hisory at the Oak Ridge National Laboratory", *Grouting: Compaction, Remediation and Testing*, Geotechnical Special Publication No.66, Ed. C. Vipulanandan, pp. 76-89.

[3.] Santagata, M.C. and Collepardi, M. (1997), "Superplasticized Microcement Grouts", 5th CANMET-ACI International Conference on *Superplasticizers and Other Chemical Admixtures in Concrete*, Rome, October 1997.

GROUTS AND GROUTING   195

GROUTS AND GROUTING 195

[4.] UNI-Italian Institute for the Standardization of Materials and Tests (1997), *Microfine Binders, Specifications and Requirements*, draft.
[5.] Mitchell, J.K. (1981), "Soil Improvement – State of the Art Report", Proc. X ICSMFE, Stockholm, Vol. 4, pp. 509-565.
[6.] Fiore, S. (1997), *Development of Design Criteria for Microcement Based Grouting* , Thesis for the Degree of Laurea in Civil Engineering, University of Ancona, Italy (in Italian).
[7.] Lowell, S. and Shields, J.E. (1984), *Powder Surface area and Porosity*, Chapman and Hall, New York.
[8.] Olson, R.A., Neubauer, C.M., and Jenings, H.M. (1997), "Damage of the Pore Structure of Hardened Portland Cement Paste by Mercury Intrusion", *J.Am.Ceram.Soc,* 80 [9], pp. 2454-58.
[9.] Santagata, M. and Collepardi, M. (1998), "Microstructural Investigation of Grouted Sands", in preparation.
[10.] De Paoli, B., Bosco, B., Granata, R., and Bruce, D.A. (1992), "Fundamental Observations on Cement Based Grouts (2): Microfine Grouts and the Cemill Process", Proc. *Grouting, Soil Improvement and Geosynthetics*, Geotechnical Special Publication No.30, ASCE, New Orleans, February 25-28, 1992, pp. 486-499.

Subject Index

Page number refers to the first page of paper

Author Index

Page number refers to the first page of paper